Student Support Materials for AQA

AS/A-level Year 1

Chemistry

Paper 1 Inorganic chemistry and relevant physical chemistry topics

Authors: Colin Chambers, Geoffrey Hallas, Andrew Maczek, David Nicholls, Rob Symonds, Stephen Whittleton

William Collins' dream of knowledge for all began with the publication of his first book in 1819.

A self-educated mill worker, he not only enriched millions of lives, but also founded a flourishing publishing house. Today, staying true to this spirit, Collins books are packed with inspiration, innovation and practical expertise. They place you at the centre of a world of possibility and give you exactly what you need to explore it.

Collins. Freedom to teach

HarperCollins Publishers
The News Building
1 London Bridge Street
London SE1 9GF

HarperCollinsPublishers
Macken House, 39/40 Mayor Street Upper,
Dublin 1,
D01 C9W8,
Ireland

Browse the complete Collins catalogue at
www.collins.co.uk

10 9 8 7 6 5

ISBN 978-0-00-818078-2

Collins® is a registered trademark of HarperCollins*Publishers* Limited

www.collins.co.uk

A catalogue record for this book is available from the British Library

Thanks to John Bentham and Graham Curtis for their contributions to the previous editions.

Commissioned by Gillian Lindsey
Edited by Alexander Rutherford
Project managed by Maheswari PonSaravanan at Jouve
Development by Tim Jackson
Copyedited and proofread by Janette Schubert
Typeset by Jouve India Private Limited
Original design by Newgen Imaging
Cover design by Angela English
Printed by Ashford Colour Press Ltd
Cover image © Shutterstock/isaravut

Contents

3.1 Physical chemistry

3.1.1 Atomic structure

3.1.1.1 Fundamental particles

Evolution of atomic structure over time

Ideas about atoms and their internal structure have evolved over thousands of years, with our understanding accelerating in the last 200 years.

Element: Aristotle (*c*350 BCE) proposed that earthly matter was made up of four elements (earth, air, wind, fire).

Atom: Democritus (*c*400 BCE) proposed that matter was made up of indivisible particles. These became known as atoms from the Greek '*atomos*', meaning 'cannot be divided'. Dalton, in his Atomic Theory (1803), resurrected the idea of atoms as being indivisible and indestructible. Atoms have different masses, and combine to form what we now call compounds.

Nucleus: When studying the penetrating effects of emissions from a radioactive source, Rutherford (1911) deduced that almost all of the mass of an atom is concentrated in its tiny centre.

Electron: Experiments on electrical discharge tubes led to the discovery of cathode rays, and then to Thomson's discovery of the electron (1897).

Proton: Rutherford, with Moseley, developed the idea that positively charged particles called protons were responsible for the positive charge on the nucleus of an atom.

Neutron: Although the presence of neutral particles in the nuclei of atoms was proposed at about the same time as the discovery of the proton, the neutron's lack of charge made its proof harder. Chadwick (1932) finally proved that neutrons do exist.

It should be appreciated that our understanding of the structure of the atom did not cease with the discovery of the neutron, and that subsequent work by, amongst others, Bohr, de Broglie, and Schrödinger, has led to our current more detailed and sophisticated knowledge.

Students of chemistry can develop all the necessary explanations for structure and bonding by assuming that an atom is made from the particles shown in Table 1.

Notes

The SI unit of charge is the coulomb (C). SI is the abbreviation used for the international system of units of measurement (Système International d'Unités).

Table 1
Properties of fundamental particles

Particle	Mass/kg	Charge/C	Relative mass	Relative charge
proton	1.673×10^{-27}	1.602×10^{-19}	1	+1
neutron	1.675×10^{-27}	0	1	0
electron	9.109×10^{-31}	1.602×10^{-19}	5.45×10^{-4}	–1

The mass of an electron is so small in comparison with the mass of either a proton or a neutron that its relative mass is often taken to be zero.

Protons, neutrons and electrons

An atom consists of electrons surrounding a small, heavy nucleus that contains protons and neutrons (except for the hydrogen atom, ^{1}H, which has only one proton and no neutrons in the nucleus).

Notes

The atomic radius of a hydrogen atom is about 10 000 times the radius of the nucleus.

3.1.1.2 Mass number and isotopes

Definition

*The **mass number**, A, of an atom is the total number of protons and neutrons in the nucleus of one atom of the element.*

Definition

*The **atomic (proton) number**, Z, of an atom is the number of protons in the nucleus of an atom.*

An atom is neutral; it has no overall charge. The charge on a proton is equal but opposite to the charge on an electron. Therefore the atomic number must also be equal to the number of electrons in a neutral atom. In an element, each atom has the same atomic number, the same number of protons and the same number of electrons. **Isotopes** of the same element consist of atoms with the same atomic number but different numbers of neutrons; their mass numbers are therefore different.

The notation for an isotope gives the mass number and the atomic number:

mass number → ^{12}
atomic number → $_{6}$C or in general $^{A}_{Z}X$

Some isotopes of hydrogen are $^{1}_{1}H$ $^{2}_{1}H$ $^{3}_{1}H$

Some isotopes of chlorine are $^{35}_{17}Cl$ $^{37}_{17}Cl$

The number of neutrons in the nucleus of an isotope can be calculated as follows:

mass number = number of protons + number of neutrons
so number of neutrons = mass number − atomic number

e.g. for $^{12}_{6}C$, the number of neutrons = 12 − 6 = 6
for $^{13}_{6}C$, the number of neutrons = 13 − 6 = 7

The chemical properties of isotopes are almost identical, because isotopes have the same number of protons and electrons. Chemical properties are dictated by the number and the arrangement of electrons. The only differences between isotopes are in physical properties, such as rates of diffusion, which depend on the mass of the particles, or in nuclear properties such as radioactivity and the ability to absorb neutrons. Different isotopes of the same element also have slightly different boiling points.

Principles of a time of flight mass spectrometer

Mass spectrometry is a powerful instrumental method of analysis. It can be used to

- find the mass and abundance of each isotope in an element, allowing its relative atomic mass to be determined
- help to identify molecules by determining their relative molecular mass.

Applications include the testing of athletes' blood or urine samples for the illegal use of performance-enhancing substances, and in space research to analyse rock and atmosphere samples.

A common form of mass spectrometry is time of flight mass spectrometry (Fig 1), in which atoms or molecules are ionised to form positive 1+ ions. The ions are then accelerated to a point where they all have the same kinetic energy. The time taken to travel a further fixed distance is used to find the mass of each ion in the sample.

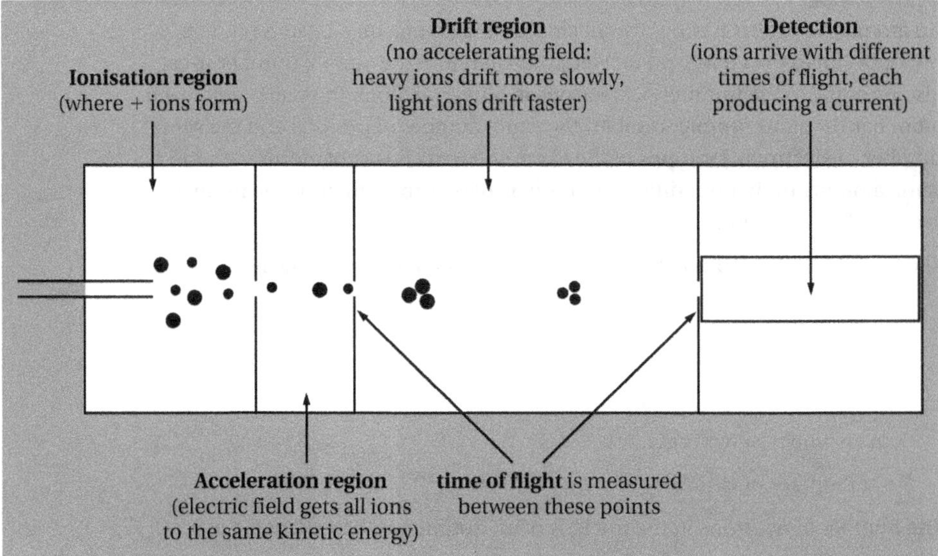

Fig 1
A time of flight mass spectrometer (simplified)

Ionisation

The neutral atoms or molecules in the sample must be turned into positive ions.

Two methods to achieve this are electron impact ionisation and electrospray ionisation.

Electron impact ionisation (also known as electron ionisation)

The sample being analysed is vaporised and high-energy electrons are fired at it. The high-energy electrons come from an 'electron gun'. The electron gun has a hot wire filament with a current running though it; the heated wire emits electrons. These electrons are accelerated by attraction to a positively charged electrode. Such high-energy electrons usually knock off only one electron from each atom or molecule in the sample forming a 1+ ion.

$M(g) + e^- \rightarrow M^+(g) + 2e^-$ (also written as $M(g) \rightarrow M^+(g) + e^-$)

The 1+ ions are then attracted towards a negative electric plate and therefore accelerate to gain the same kinetic energy regardless of mass.

This technique is used for elements and inorganic or organic molecular substances with low molecular mass. When molecules are ionised in this way, in addition to forming a molecular ion, they can break down into smaller fragments some of which are also detected in the mass spectrum.

Electrospray ionisation

The sample (X) is dissolved in a volatile solvent (e.g. water or methanol) and injected through a thin hypodermic needle to give a fine mist (aerosol). The needle tip is attached to the positive terminal of a high voltage power supply. The particles are ionised by gaining a proton (H^+) from the solvent as they leave the needle producing XH^+ ions (ions with a single positive charge and a mass of $M_r + 1$).

$$X + H^+ \rightarrow XH^+$$

The solvent evaporates away and the XH^+ ions are attracted towards the negative plate in the same way as the X^+ ions were in electron impact ionisation.

This technique is used for many substances with high molecular mass, including many biological molecules such as proteins. This procedure is known as a 'soft' ionisation technique (that is, low energy) and fragmentation rarely takes place. This outcome is an advantage because many large organic molecules subjected to electron impact ionisation fragment so readily that a molecular ion is not formed, so that the relative molecular mass cannot be determined.

Acceleration

The positive ions are accelerated using an electric field so that they all have the same kinetic energy, E_k.

$$E_k = \frac{1}{2}mv^2$$

E_k is the kinetic energy of particle (J)

m is the mass of the particle (kg)

v is the speed of the particle (m s^{-1})

So, on rearranging, the speed of each particle is given by: $v = \sqrt{\frac{2E_k}{m}}$

Given that all the particles have the same kinetic energy, the speed of each particle depends on its mass.

Lighter particles have greater speed, and heavier particles have a lesser speed.

Ion drift

The positive ions travel through a hole in the negatively charged plate into a tube.

The time of flight of each particle through this flight tube depends on its speed ($t = d/v$) and therefore on its mass.

Essential Notes

All the 1+ ions have the same kinetic energy.

Notes

The study of fragments is not required within this specification.

Essential Notes

Particles gain a proton from the solvent so X becomes XH^+.

Essential Notes

Soft ionisation impedes the fragmentation of the protonated molecular ion XH^+.

Notes

You will be given this equation if you are expected to use it in an exam.

Essential Notes

The mass of the charged particle determines its speed and hence its time of flight.

The time of flight along the flight tube is given by the following expression where d is the length of the tube:

$$t = d\sqrt{\frac{m}{2E_k}}$$

t is the time of flight (s)

E_k is the kinetic energy of particle (J)

m is the mass of the particle (kg)

d is the length of flight tube (m)

This equation shows that the time of flight is proportional to the square root of the mass of the ions. Therefore lighter ions travel fast and reach the detector in less time and the heavier particles travel more slowly and take longer to reach the detector.

For example, ions of the three isotopes of magnesium ($^{24}Mg^+$, $^{25}Mg^+$, $^{26}Mg^+$) will travel at different speeds through the flight tube and separate, with the lightest ion ($^{24}Mg^+$) reaching the detector first.

Example

Typical time of flight calculation:

$^{26}Mg^+$ ion has relative mass = 26

The actual mass of the ion = $26/L$ g = $26 \times 10^{-3}/L$ kg, where L = the Avogadro constant, 6.022×10^{23} mol^{-1}

Therefore $m = 26 \times 10^{-3}/6.022 \times 10^{23} = 4.35 \times 10^{-26}$ kg

So if $E_k = 2.175 \times 10^{-16}$ J, and $d = 0.6$ m

Then

$$t = 0.6\sqrt{\frac{4.35 \times 10^{-26}}{2 \times 2.175 \times 10^{-16}}}$$

$$= 0.6 \times 10^{-5}\ \text{s} = \mathbf{6 \times 10^{-6}\ s}$$

Essential Notes

In mass spectrometry, m/z is known as the mass-to-charge ratio, where m is the relative molecular, atomic or fragment mass and z is the charge on the ion. Because the ionisation process is designed to produce only 1+ ions, m/z values in a printout are a direct measure of relative mass.

Detection

At the end of the drift tube, the positive ions strike a negatively charged electric plate. When they hit this detector plate, the positive ions are neutralised by gaining electrons from the plate. This process generates a flow of electrons and hence an electric current that is then amplified to produce a signal on a computer.

The relative intensity of the peak in the resulting mass spectrum produced by an ion with a particular m/z value (mass-to-charge ratio) is proportional to the magnitude of the amplified current. This current is proportional to the number of ions hitting the plate. Therefore the current and hence the peak height give a measure of the abundance of the ion.

A typical printout from a sample of chlorine in a mass spectrometer that uses electron impact ionisation is shown in Fig 2.

- Molecules ionised using electron impact ionisation give rise to a peak with a maximum value of $m/z = M_r$.
- Molecules ionised using electrospray ionisation give rise to a peak with a maximum value of $m/z = (M_r + 1)$.

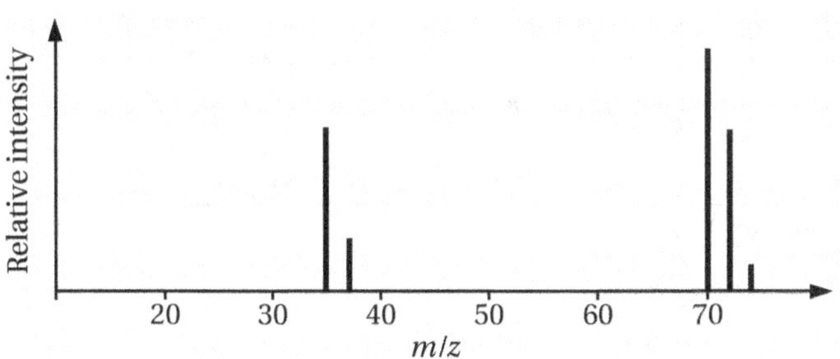

Fig 2
The mass spectrum of chlorine

Essential Notes

The mass spectrometer is sensitive enough to distinguish clearly between isotopes.

Notes

For simplicity, the atomic number of isotopes is often omitted in symbols such as ^{35}Cl.

The peak at $m/z = 35$ represents the $^{35}Cl^+$ ion. The ratio of peak heights at $m/z = 35$ and 37 is 3 : 1. The peak heights for the Cl_2^+ ions are in the ratio 9 : 6 : 1. This ratio represents the proportions of Cl_2^+ ions with $m/z = 70$, 72 and 74, respectively.

The peaks in the mass spectrum can be assigned as in Table 2.

Table 2
Assignment of peaks in the mass spectrum of chlorine

m/z	35	37	70	72	74
Ion from	^{35}Cl	^{37}Cl	$^{35}Cl — ^{35}Cl$	$^{37}Cl — ^{35}Cl$	$^{37}Cl — ^{37}Cl$

The **relative atomic mass** (A_r) of an isotopic mixture can be calculated by using information from a mass spectrum. The spectrum above shows that the relative proportions of isotopes in that sample of chlorine are:

$^{35}Cl : ^{37}Cl \quad = \quad 3 : 1$

so $\frac{3}{4}$ of the sample is ^{35}Cl and $\frac{1}{4}$ of the sample is ^{37}Cl

so $A_r = \frac{3}{4} \times 35 + \frac{1}{4} \times 37 = 35.5$

The *effective relative atomic mass* of an element is determined by calculating the weighted mean of the individual relative atomic masses of the isotopes. The *abundance* of the different isotopes is found from a mass spectrum.

Essential Notes

A mass spectrometer can be used to identify elements from the *m/z* values of their isotopes.

Elements that have an extra-terrestrial origin (e.g. those found in meteorites) often have a different ratio of isotopes compared with the same element on planet Earth.

The **relative molecular mass** (M_r) of a substance can also be determined from its mass spectrum. When a sample, X, is introduced into a mass spectrometer the peaks near the maximum value of m/z correspond to molecular ions, X^+, made up from the various isotopic mixtures. The relative molecular mass of the substance can be calculated from these peaks.

Notes

Isotopic masses are quoted as integer values but these are not the exact mass values.

3.1.1.3 Electron configuration

Electron arrangement in atoms and ions

It was originally considered that the maximum number of electrons that could be accommodated in the outside layer of an atom was eight and that these electrons occupied a circular or spherical orbit. The elements from helium to krypton were thought to be inert because their outer orbits were full.

This model of atomic structure has been replaced by one that is able to account for observed facts, such as the reactions of fluorine with these gases which are now referred to as 'noble' rather than 'inert'. In this later model, the electrons in atoms are arranged into main (or principal) **energy levels**, which are numbered. Level 1 contains electrons which are closest to the nucleus. Within levels there are sub-levels designated s, p, d, f. The maximum number of sub-levels is different for each level and is shown in Table 3.

Essential Notes

For definitions of the terms A_r and M_r see this book, section 3.1.2.1.

Each sub-level consists of **orbitals**. Each orbital can hold a maximum of two electrons which have opposite spin. The number of orbitals and the maximum number of electrons which can be accommodated in each sub-level are shown in Table 4.

Table 3
Types of sub-level in each electron energy level (shell)

Main (principal) level	1	2	3	4
Sub-levels in that level	s	s, p	s, p, d	s, p, d, f

Table 4
Number of orbitals and maximum number of electrons in each sub-level

Sub-level	s	p	d	f
Number of orbitals in sub-level	1	3	5	7
Maximum number of electrons	2	6	10	14

Notes

Main (principal) energy levels for electrons are sometimes referred to as shells.

Owing to differences in shielding from the nucleus, different sub-levels within a level have slightly different energies. A typical energy-level diagram for an atom is shown in Fig 3.

Fig 3
A typical energy-level diagram

— — — (4p) — — — — — (3d)
level 4 — (4s) — — — (3p)
level 3 — (3s)
— — — (2p)
level 2 — (2s)
level 1 — (1s)
each orbital can hold up to 2 electrons

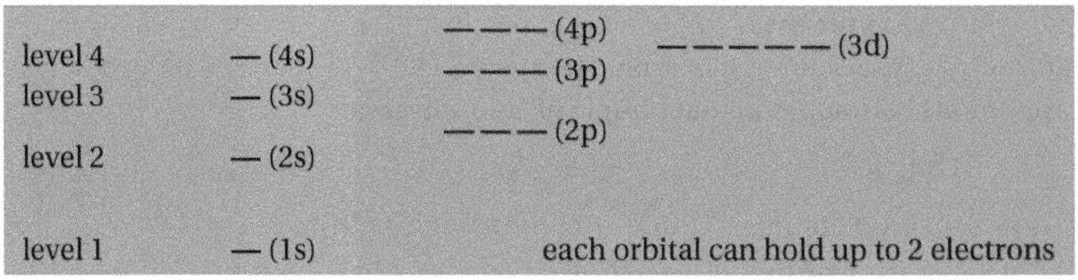

Level 2 sub-levels all lie below level 3 sub-levels but, in atoms, sub-level 3d is higher in energy than sub-level 4s.

The electron configurations of atoms of elements can be deduced from this diagram. Typical configurations are:

- Li $1s^2 2s^1$
- F $1s^2 2s^2 2p^5$
- Fe $1s^2 2s^2 2p^6 3s^2 3p^6 4s^2 3d^6$

The electron configurations of ions can be deduced from the configuration of the neutral atom by adding or removing electrons. A complication is that, for transition-metal ions, the 3d sub-level is lower in energy than the 4s, so that the 4s electrons are removed first:

- Li^+ $1s^2$
- F^- $1s^2 2s^2 2p^6$
- Fe^{3+} $1s^2 2s^2 2p^6 3s^2 3p^6 3d^5$

Essential Notes

If electrons are regarded as clouds rather than as particles, the electron cloud has a characteristic shape for each type of orbital.

Essential Notes

An s orbital is an electron cloud with spherical symmetry.

The diagrams of p orbitals are shown superimposed on 3 dimensional *x*, *y*, *z* axes. Each p orbital has twin lobes.

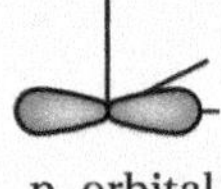

p_x orbital

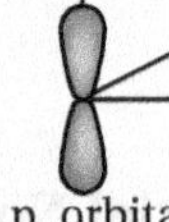

p_y orbital

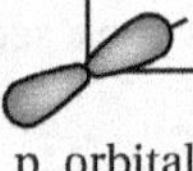

p_z orbital

(d and f orbitals have more complicated shapes.)

First ionisation energy and electron arrangements

Definition

*The **first ionisation energy** of an element is defined as the enthalpy change for the removal of one mole of electrons from one mole of atoms of the element in the gas phase:*

$X(g) \rightarrow X^+(g) + e^-$

The first ionisation energies of the Group 2 metals (Be–Ba) vary as shown in Fig 4.

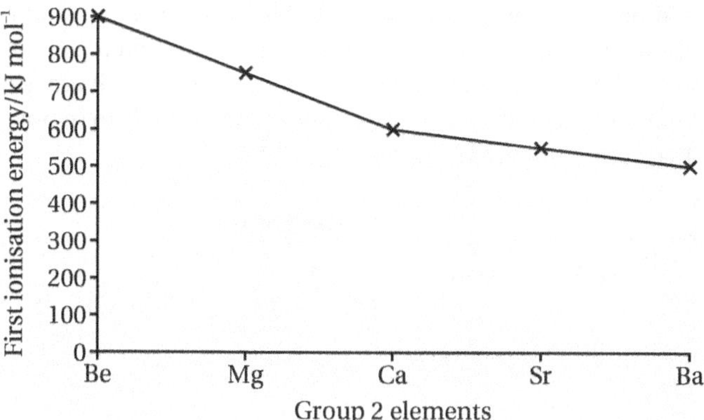

Fig 4
First ionisation energies of the Group 2 elements

Notes

Electron configurations are sometimes abbreviated by giving only the electrons beyond the previous noble gas. For example

Fe [Ar] $4s^2 3d^6$

Fe^{3+} [Ar] $3d^5$

For **neutral atoms** the 3d sub-level is higher in energy than the 4s. For **ions**, this is no longer true. The 3d sub-level becomes lower in energy than the 4s.

There is a successive decrease in first ionisation energy from beryllium to barium. Magnesium has a lower first ionisation energy than beryllium because its outer electron is in a 3s sub-level rather than a 2s sub-level. The 3s sub-level is higher in energy than the 2s sub-level. The 3s electron is further from the nucleus and is more shielded from the nucleus by inner electrons. Thus, the 3s electron is more easily removed. This trend in ionisation energies is evidence for the electrons of atoms being organised in levels. A similar decrease in ionisation energy occurs down each group in the Periodic Table.

Notes

Note that the decrease in ionisation energy down the group is small in comparison with the absolute magnitude of the ionisation energies.

The first ionisation energies of the elements from neon to potassium vary as shown in Fig 5.

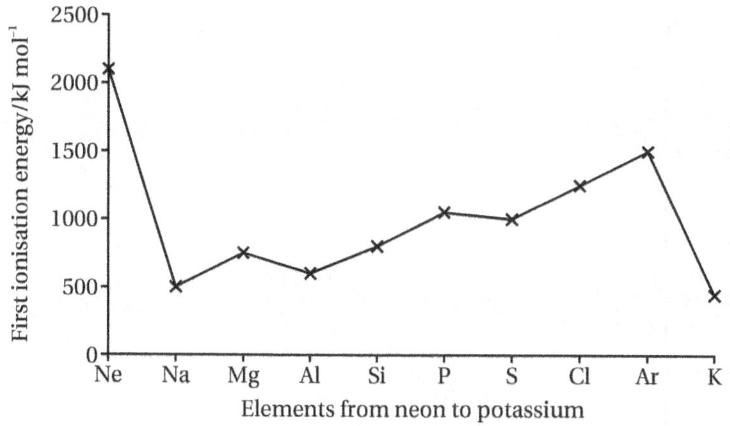

Fig 5
First ionisation energies of the elements neon to potassium

Notes

The first ionisation energy of neon is greater than that of sodium because the outermost electron in sodium is in a main level which is further from the nucleus and more shielded. This big difference in ionisation energy from neon to sodium is strong evidence for the existence of main (principal) electron energy levels.

There is a general increase in ionisation energy across Period 3 (sodium to argon). Across the period from Na (11 protons) to Ar (18 protons) the nuclear charge in each element increases. As a result, the electrons are attracted more strongly to the nucleus and it takes more energy to remove one from the atom.

There is a fall in ionisation energy from magnesium to aluminium because the outer electron in Al (configuration $1s^2 2s^2 2p^6 3s^2 3p^1$) is in a p sub-level. The p sub-level electron is higher in energy than the outer electron in Mg ($1s^2 2s^2 2p^6 3s^2$) which is in an s sub-level.

The fall in ionisation energy from phosphorus to sulfur can be explained by considering their electronic arrangements (see Fig 6).

Essential Notes

Electrons have a property called spin. A spinning electron can be represented by an arrow. Electrons can only spin in one of two directions and are shown by an up or a down arrow, i.e. ↑ or ↓. The arrows represent the magnetic field produced by the spinning electron.

The 3p electrons in phosphorus are unpaired. If there are several empty sub-levels all of the same energy, electrons will organise themselves so that they remain unpaired and occupy as many sub-levels as possible. In sulfur the fourth 3p electron is paired. There is some repulsion between paired electrons in the same sub-level, which increases their energy. Therefore it is easier to remove one of these paired 3p electrons from sulfur than it is to remove an unpaired 3p electron from phosphorus.

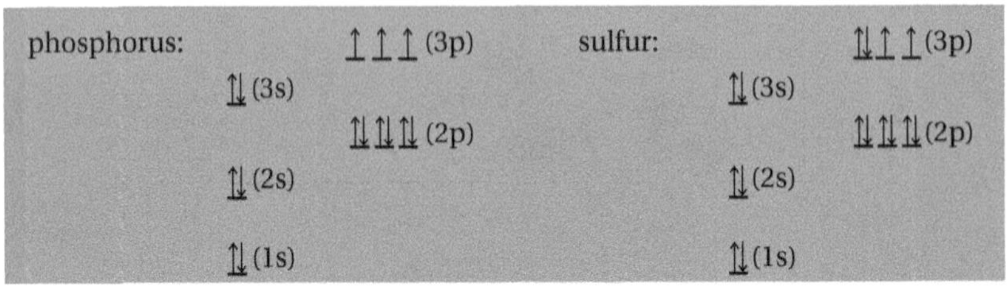

Fig 6
Energy levels for phosphorus and sulfur

phosphorus:		sulfur:	
	↑ ↑ ↑ (3p)		↑↓ ↑ ↑ (3p)
↑↓ (3s)		↑↓ (3s)	
	↑↓ ↑↓ ↑↓ (2p)		↑↓ ↑↓ ↑↓ (2p)
↑↓ (2s)		↑↓ (2s)	
↑↓ (1s)		↑↓ (1s)	

Notes

This variation across Period 3 is regarded as evidence for the existence of electronic sub-levels.

Subsequent ionisation energies and their relationship to electron shells

The second and third ionisation energies of an element X are the **enthalpy changes** for the reactions:

$$X^+(g) \rightarrow X^{2+}(g) + e^-$$
$$X^{2+}(g) \rightarrow X^{3+}(g) + e^-$$

Definition

***Second ionisation energy** is the enthalpy change for the removal of one mole of electrons from one mole of unipositive ions in the gas phase:*

$$X^+(g) \rightarrow X^{2+}(g) + e^-$$

Successive ionisation energies can provide a very useful guide to the number of electrons in the outside shell (electron energy level) of an element. For example, the successive ionisation energies for aluminium vary as Fig 7 shows.

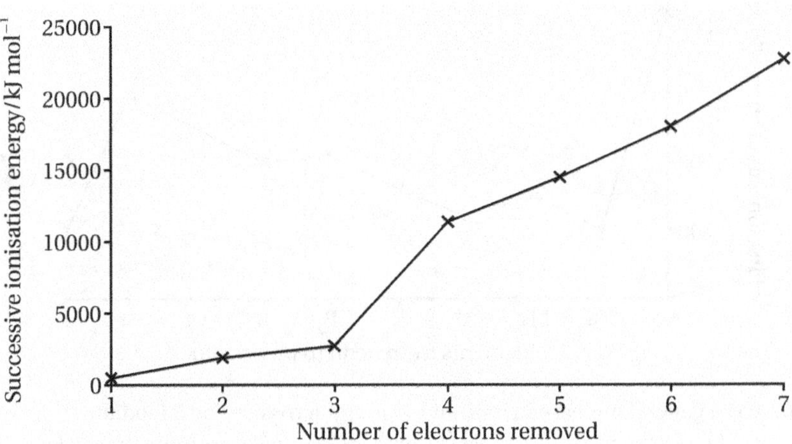

Fig 7
Successive ionisation energies for aluminium

There is a big jump after removal of the third electron because the next electron must be removed from an inner shell. The graph in Fig 7 shows that aluminium has three electrons in its outside shell and is therefore in Group 3.

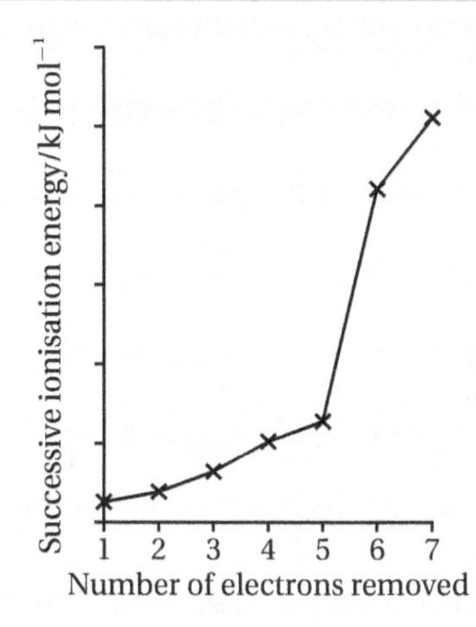

Example

In which group in the Periodic Table is this element to be found?

Answer

The large jump after the fifth electron shows that this element is in Group 5.

3.1.2 Amount of substance

3.1.2.1 Relative atomic mass and relative molecular mass

The isotope ^{12}C is the standard for relative mass:

$$\text{relative atomic mass } (A_r) = \frac{\text{average mass per atom of an element}}{\frac{1}{12} \times \text{mass of one atom of } ^{12}C}$$

$$\text{relative molecular mass } (M_r) = \frac{\text{average mass of a molecule}}{\frac{1}{12} \times \text{mass of one atom of } ^{12}C}$$

Notes

The average mass must be used to allow for the presence of isotopes.

For compounds that are not molecules, relative formula mass is used:

$$\text{relative formula mass } (M_r) = \frac{\text{average mass of an 'entity'}}{\frac{1}{12} \times \text{mass of one atom of } ^{12}C}$$

Essential Notes

An 'entity' is a 'formula unit'.

3.1.2.2 The mole and the Avogadro constant

The **Avogadro constant** is a quantity and is given the symbol L:

$$L = 6.022 \times 10^{23} \text{ mol}^{-1}$$

Essential Notes

The symbol for the unit of the mole is mol. The SI units of the Avogadro constant are mol^{-1}. It is often helpful to think of this as the number of particles that make up one mole of these particles.

The name **mole** is given to the amount of substance:

1 mol of particles/entities is 6.022×10^{23} particles/entities

For example:

1 mol of lithium atoms has a mass of 1.152×10^{-23} g

1 mol of lithium atoms contains 6.022×10^{-23} lithium atoms

therefore, 1 mol of lithium atoms has a total mass of

$$(1.152 \times 10^{-23}) \times 6.022 \times 10^{23} = 6.937 \text{ g}$$

1 mol of ^{12}C has a mass of precisely 12.000 g because ^{12}C is the standard.

Essential Notes

The mass of one atom of an element is very tiny. The mass of L atoms of an element is a recognisable number of grams.

1 mol of CH_4 molecules has a mass of approximately 16.0 g. It is calculated by adding up the individual values of the relative atomic masses:

Notes

This degree of accuracy – to one decimal place – is sufficient for most chemical purposes.

$1 \times C = 12.0$

$4 \times H = 4 \times 1.0 = 4.0$

$\text{total mass} = 12.0 + 4.0 = 16.0$

This mass is approximate for two reasons:

- the relative atomic mass of ^1_1H is 1.0078
- carbon and hydrogen occur naturally as isotopic mixtures.

Different isotopes of carbon and hydrogen, such as ^{13}C and ^2H, occur naturally, so a few of the methane molecules have a mass which is greater than the mass of a $^{12}C^1H_4$ molecule.

A chemical equation usually implies quantities in moles. For example:

CH_4	+	$2O_2$	→	CO_2	+	$2H_2O$
1 mol of methane molecules		2 mol of oxygen molecules		1 mol of carbon dioxide molecules		2 mol of water molecules

The mole can be applied to electrons, atoms, molecules, ions, formulas and equations.

The **concentration** of a solution is a quantitative expression with units of mol dm^{-3}:

$$\text{concentration} = \frac{\text{amount in moles of solute}}{\text{volume of solution in dm}^3}$$

Notes

Problems with units? Think of a metre rule:

1 m = 10 dm = 100 cm

$1\ m^3 = 10^3\ dm^3 = (100)^3\ cm^3 = 10^6\ cm^3$

A 1.0 mol dm^{-3} solution contains 1 mol of a substance which has been dissolved in enough water to make 1 dm^3 of solution. For example:

a 1.0 mol dm^{-3} solution of Na_2SO_4 contains:

$(23.0 \times 2) + 32.0 + (16.0 \times 4) = 142.0$ g of Na_2SO_4 in 1 dm^3 of solution

Three of the most useful methods for calculating the amount in moles of a substance are as follows:

1 *For a known mass of substance:*

$$\text{amount in moles} = \frac{\text{mass}}{M_r} = \frac{\text{mass}}{\text{mass of 1 mol}} = \frac{\text{mass in g}}{M_r \text{ expressed in g}}$$

2 *For solutes in a solution, if the volume of the solution is known:*

$$\text{amount in moles of solute} = \text{volume of solution in dm}^3 \times \text{concentration}$$

$$= \frac{\text{volume of solution in cm}^3}{1000} \times \text{concentration}$$

3 *For gases:*

Essential Notes

This equation is discussed in more depth in the next section.

$$\text{amount in moles} = \frac{pV}{RT}$$

3.1.2.3 The ideal gas equation

An **ideal gas** obeys the assumptions of the kinetic theory of gases. According to this theory, ideal gas particles (molecules or free atoms) are treated as hard spheres of negligible size which move with rapid random motion and experience no intermolecular forces.

The ideal gas equation is:

$$pV = nRT$$

p is the pressure of the gas in Pa

V is the volume of the gas in m^3

n is the amount in moles of gaseous particles

R is the gas constant ($8.31\ J\ K^{-1}\ mol^{-1}$)

T is the temperature in kelvin (add 273 to the temperature in °C)

This equation can be used to find the amount in moles (n) of a gaseous substance. For example, in 200 cm^3 of CH_4 at 25 °C and 100 kPa the amount in moles of methane is:

$$n = \frac{pV}{RT} = \frac{100\,000 \times 200 \times 10^{-6}}{8.31 \times 298} = 0.00808\ \text{mol}$$

$$= 8.08 \times 10^{-3}\ \text{mol (to three significant figures)}$$

If the mass and the amount in moles of a sample are known, it is possible to calculate the relative molecular mass (M_r). The mass of the sample of methane above is 0.129 g. Hence:

$$M_r = \frac{\text{mass}}{\text{amount in moles}} = \frac{0.129}{0.00808} = 16.0$$

Essential Notes

One pascal (Pa) is one newton per square metre ($N\ m^{-2}$)

Essential Notes

100 kPa = 100 000 Pa

200 cm^3 = $200 \times 10^{-6}\ m^3$

25 °C = 298 K

Notes

This is the basis of experiments to determine M_r by measuring the mass of a given volume of gas or vapour at a known temperature and pressure (gas syringe or bulb experiments).

3.1.2.4 Empirical and molecular formula

The **empirical formula** is the formula which represents the *simplest ratio* of atoms of each element in a compound.

The **molecular formula** gives the *actual number* of atoms of each element in a molecule (or the amount in moles of each type of atom in 1 mol of the compound).

Calculation of empirical formulas

The empirical formula of a compound can be calculated from data which give the percentage composition by mass of each element in the compound.

Calculation of molecular formulas

The molecular formula can be deduced from the empirical formula if the relative molecular mass is known. A value for M_r can be determined as shown using the ideal gas equation or from a mass spectrum. If a compound with empirical formula CH_2O has $M_r = 180$, the molecular formula can be calculated as shown at the end of the example on the following page.

Example

A compound containing carbon, hydrogen and oxygen gave, after elemental analysis, the following percentages by mass:
C 40% and H 6.7%.

The percentage of oxygen is often calculated by difference. In this case, the percentage of oxygen = 100 − (40 + 6.7) = 53.3%.

The empirical formula can be calculated as follows.

Assume that there are 100 g of the compound, then the masses of the elements are:

C 40 g; H 6.7 g; O 53.3 g

The amount in moles of each element is calculated as follows:

$$\text{carbon: } \frac{\text{mass}}{A_r} = \frac{40}{12.0} = 3.3 \quad \text{hydrogen: } \frac{6.7}{1.0} = 6.7 \quad \text{oxygen: } \frac{53.3}{16.0} = 3.3$$

These amounts in moles can be expressed as a simple ratio by dividing through by the smallest number:

$$\text{ratio of moles of } C:H:O = 3.3:6.7:3.3$$

$$= \frac{3.3}{3.3}:\frac{6.7}{3.3}:\frac{3.3}{3.3}$$

$$= 1:2:1$$

Therefore the empirical formula is CH_2O

The empirical formula mass of CH_2O is 12.0 + 2.0 + 16.0 = 30.0

The ratio of M_r : empirical formula mass = 180 : 30.0 = 6 : 1

Therefore, in comparison with the empirical formula, the molecular formula must contain 6 times the number of atoms.

Therefore the molecular formula is $6 \times CH_2O = C_6H_{12}O_6$

Notes

The molecular formula is always a whole number times the empirical formula.

3.1.2.5 Balanced equations and associated calculations

A **full equation** is a balanced symbol equation with the formulas of all reagents on the left-hand side and the formulas of all the products on the right-hand side.

An **ionic equation** is a simplified version of a balanced symbol equation, showing only the ions which are actively involved in the reaction. Ions which do not take part in the reaction (spectator ions) are eliminated. The resulting ionic equation is no longer specific to a single reaction but is now a generalised expression of the essential chemistry.

State symbols are often included in both full and ionic equations. These are letters, in brackets, which are added after a formula to indicate the state the substances are in:

(s) = solid (l) = liquid (g) = gas

(aq) = aqueous solution/dissolved in water

Essential Notes

For ionic equations the charges on the ions must also balance.

Thus $Fe^{3+} + Zn \rightarrow Fe + Zn^{2+}$ is not a balanced equation but

$2Fe^{3+} + 3Zn \rightarrow 2Fe + 3Zn^{2+}$

is balanced (6 positve charges on each side).

Balancing equations

Balanced equations must have the same number of atoms of each element on the left-hand side and on the right-hand side of the 'arrow'.

To balance equations, work through these steps.

1 Write the equation, then pick one element and see if the number of atoms of that element is equal on both sides of the arrow.

2 If the equation needs balancing, write the necessary number in front of the appropriate formula or symbol to make that element balance.

3 Move on to each new element and balance it in turn.

4 Check for fractions and multiply them out.

Example

Consider the reaction between sulfuric acid and sodium hydroxide.

The balanced symbol equation is:

$H_2SO_4(aq) + 2NaOH(aq) \rightarrow Na_2SO_4(aq) + 2H_2O(l)$

Showing separate ions this becomes:

$2H^+(aq) + SO_4^{2-}(aq) + 2Na^+(aq) + 2OH^-(aq) \rightarrow 2Na^+(aq) + SO_4^{2-}(aq) + 2H_2O(l)$

Water is covalent, so is not shown as ions.

Sulfate ions and sodium ions, the spectator ions, appear on both sides so are cancelled out to leave:

$2H^+(aq) + 2OH^-(aq) \rightarrow 2H_2O(l)$

This simplifies to give:

$H^+(aq) + OH^-(aq) \rightarrow H_2O(l)$

This ionic equation now represents the reaction of any acid with any hydroxide in solution.

Example

Consider the unbalanced equation:

$Al + NaOH \rightarrow Na_3AlO_3 + H_2$

Taking each element in turn:

Al There is one atom (or 1 mol of atoms) on each side, so Al balances.

Na There is one Na on the left and three on the right – the equation is unbalanced.

Therefore use 3NaOH and the equation is now

$Al + 3NaOH \rightarrow Na_3AlO_3 + H_2$

O There are now three Os on each side, so O balances.

H There are three Hs on the left and two on the right. Using $\frac{3}{2}H_2$ on the right-hand side balances the equation:

$Al + 3NaOH \rightarrow Na_3AlO_3 + \frac{3}{2}H_2$

This equation is balanced, but it is better multiplied by 2 to avoid the fraction $\frac{3}{2}$:

$2Al + 6NaOH \rightarrow 2Na_3AlO_3 + 3H_2$

Calculating reacting masses and reacting volumes of gases

This skill is again best learned from examples.

Notes

In calculations of this type the answer is usually expressed to three significant figures. It may be necessary to carry more precise numbers through the calculation, but the answer should be rounded.

Example

Consider the following equation:

$$2HCl + Na_2SO_3 \rightarrow 2NaCl + H_2O + SO_2$$

If 1.00 g of Na_2SO_3 is reacted with an excess of HCl, calculate:

(i) the mass of NaCl produced by complete reaction

(ii) the volume of SO_2 gas produced at 25°C and 100 kPa pressure.

Answer

Calculations like this almost always involve, as an intermediate step, working out amounts in moles. The answers can be deduced as follows:

(i) M_r for Na_2SO_3 is $(2 \times 23.0) + 32.0 + (3 \times 16.0) = 126.0$

$$\text{moles of } Na_2SO_3 \text{ used} = \frac{\text{mass}}{M_r} = \frac{1.00}{126.0} = 0.00794$$

From the equation, moles of NaCl $= 2 \times$ moles of Na_2SO_3
$= 2 \times 0.00794$
$= 0.0159$

M_r for NaCl $= 23.0 + 35.5 = 58.5$

mass of NaCl $=$ moles $\times M_r$
$= 0.0159 \times 58.5 = 0.929$ g

(ii) From the equation, moles of SO_2 = moles of Na_2SO_3 = 0.00794

$$\text{volume } V = \frac{nRT}{p} = \frac{0.00794 \times 8.31 \times 298}{100\,000} = 1.97 \times 10^{-4}\ \text{m}^3 = 197\ \text{cm}^3$$

Notes

Remember

$1\ \text{m}^3 = 1 \times 10^6\ \text{cm}^3$

$1\ \text{cm}^3 = 1 \times 10^{-6}\ \text{m}^3$

Calculating concentrations and volumes of aqueous reagents

Work through this example of a typical problem.

Notes

The general way to progress through calculations like this is

- calculate the moles of known substance
- use the equation for the reaction and the moles of known substance to state the moles of unknown substance
- proceed to the answer.

Example

25.0 cm^3 of 0.102 mol dm^{-3} NaOH are exactly neutralised by a solution of 0.0830 mol dm^{-3} H_2SO_4:

$$H_2SO_4(aq) + 2NaOH(aq) \rightarrow Na_2SO_4(aq) + 2H_2O(l)$$

Calculate:

(i) the volume of sulfuric acid required for the neutralisation

(ii) the concentration of sodium sulfate in the resulting solution.

Answer

(i) moles of NaOH = volume (in dm^3) $\times$ concentration

$$= \frac{\text{volume (in cm}^3\text{)}}{1000} \times \text{concentration}$$

$$= \frac{25.0}{1000} \times 0.102$$

$$= 0.00255 \text{ mol}$$

From the equation:

$$\text{moles of } H_2SO_4 = \frac{1}{2} \times \text{moles of NaOH}$$

$$= \frac{0.00255}{2}$$

$$= 0.001275 \text{ mol}$$

amount in moles of H_2SO_4 = volume in dm^3 × concentration

$$\text{therefore volume of } H_2SO_4(aq) = \frac{\text{amount in moles}}{\text{concentration}}$$

$$= \frac{0.001275}{0.0830}$$

$$= 0.0154 \text{ dm}^3$$

$$= 15.4 \text{ cm}^3$$

(ii) moles of Na_2SO_4 produced = moles of H_2SO_4 used = 0.001275 mol

Ignoring the small amount of water produced in the reaction:

total volume of final solution = 25.0 + 15.4 = 40.4 cm^3

$= 40.4 \times 10^{-3}$ dm^3

$$\text{concentration of } Na_2SO_4(aq) = \frac{\text{amount in moles}}{\text{volume in dm}^3}$$

$$= \frac{0.001275}{40.4 \times 10^{-3}} = 0.0316 \text{ mol dm}^{-3}$$

Notes

Remember

concentration

$$= \frac{\text{moles of solute}}{\text{volume of solution in dm}^3}$$

Also $1 \text{ dm}^3 = 1 \times 10^3 \text{ cm}^3$

$1 \text{ cm}^3 = 1 \times 10^{-3} \text{ dm}^3$

Notes

The making up of a volumetric solution and the carrying out of an acid–base titration is a required practical activity.

Percentage yield

Percentage yield is a practical measure of the efficiency of a reaction. It takes into account reactions that do not go to completion. It can only be calculated from experimental data.

$$\text{percentage yield} = \frac{\text{actual mass of product}}{\text{maximum theoretical mass of product}} \times 100$$

Example

Consider the following equation for the production of dichloromethane (CH_2Cl_2):

$$CH_4 + 2Cl_2 \rightarrow CH_2Cl_2 + 2HCl$$

In an experiment, 21.3 g of CH_2Cl_2 were produced when 8.00 g of methane were reacted with an excess of chlorine.

$$\text{amount in moles of methane} = \frac{8.00}{16.0} = 0.500$$

Maximum amount in moles of CH_2Cl_2 that can be formed from 0.500 mol of CH_4 = 0.500 mol

Maximum mass of CH_2Cl_2 that can be formed

= amount in moles × M_r = 0.500 × 85.0 = 42.5 g

Actual mass of CH_2Cl_2 formed = 21.3 g

$$\text{yield} = \frac{\text{actual mass of } CH_2Cl_2 \times 100}{\text{maximum theoretical mass of product}}$$

$$= \frac{21.3}{42.5} \times 100 = 50.1\%$$

This answer suggests that the reaction did not go to completion or that some of the methane was converted into by-products (or a combination of the two).

Percentage atom economy

The **percentage atom economy** is a measure of how much of a desired product in a reaction is formed from the reactants. It is a theoretical quantity calculated from a balanced equation.

$$\text{percentage atom economy} = \frac{\text{mass of desired product}}{\text{total mass of reactants}} \times 100$$

Example

Consider the following equation for the production of dichloromethane (CH_2Cl_2):

$$CH_4 + 2Cl_2 \rightarrow CH_2Cl_2 + 2HCl$$

$$\text{percentage atom economy} = \frac{\text{mass of one mole of } CH_2Cl_2 \times 100}{(\text{mass of one mole of } CH_4 + \text{mass of two moles of chlorine})}$$

$$= \frac{85.0 \times 100}{(16.0 + 142.0)} = 53.8\%$$

This answer shows that (100 - 53.8) = 46.2% of the mass of reactants is converted into a co-product other than the desired product.

Economic, ethical and environmental advantages of high atom economy

Unlike percentage yield improvement, which focuses only on the amount of product formed in a reaction, high atom economy, developed from the principles of 'green' or sustainable chemistry, focuses on economic, ethical and environmental issues.

Adoption of high atom economy processes reduces waste products and so reduces both the cost of hazardous waste treatment and potential damage caused by its release into the environment. In addition, high atom economy synthetic routes are less wasteful of natural resources.

If the quest for higher atom economy processes also incorporates the use of less toxic starting materials and safer solvents, greater use of renewable resources, the development of more efficient catalysts, and less energy usage, there are further indirect advantages.

One example of a successful increase in atom economy is the synthesis of Ibuprofen. The original manufacturing process had an atom economy of 40% but switching to an alternative synthetic route has increased this to 77%.

3.1.3 Bonding

3.1.3.1 Ionic bonding

Positive **ions** are formed when atoms lose electrons:

$Li\ (1s^2 2s^1) \rightarrow Li^+\ (1s^2) + e$

Negative ions are formed when atoms gain electrons:

$O\ (1s^2 2s^2 2p^4) + 2e \rightarrow O^{2-}\ (1s^2 2s^2 2p^6)$

Compounds which contain ionic bonds are solids at room temperature. In solids, **ionic bonds** never exist in isolation. They form part of a **giant ionic lattice** where each positive ion is attracted by negative ions which surround the positive ion in a regular array. For example, the structure of NaCl in two dimensions is as shown in Fig 8. The positive ions are electrostatically attracted to the negative ions.

Essential Notes

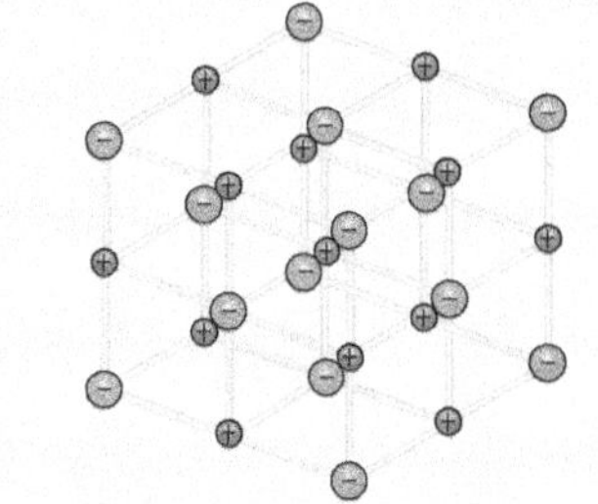

The sodium chloride lattice can also be drawn like this. The ions in this diagram are not drawn to be *space-filling.*

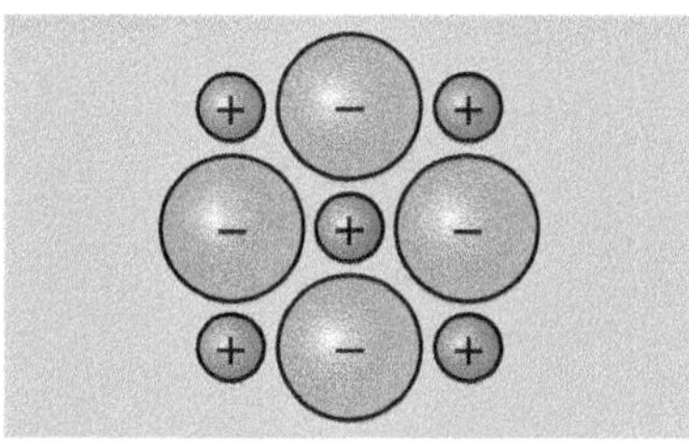

Fig 8
A slice through a three-dimensional ionic lattice of NaCl

Ionic compounds conduct electricity when molten or in aqueous solution. The ions are fixed in position in the solid phase, but in the liquid phase, or in solution, they are free to move and carry the current.

Predicting the charge on a simple ion

Simple ions consist of one type of element only. The position of the element in the Periodic Table can be used to predict the charge on the ion.

Elements in Groups 1 to 3 form positive ions by electron loss.

For these, charge on the ion = the group number or the number of outermost electrons.

For example, aluminium is in Group 3, with 3 outermost electrons, so its ion is Al^{3+}.

Elements in Groups 6 and 7 form negative ions by electron gain.

For these, charge on the ion = 8 − the group number or

8 − the number of outermost electrons.

For example, sulfur is in Group 6, with 6 outermost electrons, so its ion is S^{2-}.

Note also that these negative ions have names ending in -ide, e.g. sulfide, chloride.

Elements in Groups 4 and 5 generally form covalently bonded compounds. The only known compounds formed by elements of Group 0, such as XeF_4, are covalently bonded.

Compound ions contain 2 or more elements. Some of the commonest are:

sulfate, SO_4^{2-} hydroxide, OH^-

nitrate, NO_3^- carbonate, CO_3^{2-}

ammonium, NH_4^+

To construct the formula of an ionic compound, the positive and negative ions must combine in a ratio that cancels out the charges.

For example, potassium oxide:

K is in Group 1, so forms the ion K^+. O is in Group 6, so forms the ion O^{2-}.

Two K^+ ions are needed to cancel out the –2 charge on the O^{2-}.

So, the formula of potassium oxide is K_2O.

For example, ammonium sulfate:

Ammonium is a compound ion with formula NH_4^+. Sulfate is a compound ion with formula SO_4^{2-}.

Two NH_4^+ ions are needed to cancel out the –2 charge on the SO_4^{2-}.

So, the formula of ammonium sulfate is $(NH_4)_2SO_4$.

Note the use of brackets here for the ammonium ion. Whenever more than one compound ion is needed in a formula, the multiple is shown by a subscript. This subscript indicates that the entire contents of the brackets are multiplied. So, in the formula of ammonium sulfate, we have 2 × N atoms, 8 × H atoms, 1 × S atom and 4 × O atoms.

3.1.3.2 Nature of covalent and dative covalent bonds

Notes

The shared pair of electrons in a covalent bond is usually represented by a line like this:

H—F

Unlike ionic bonds, **covalent bonds** can exist in isolation in single molecules.

A single covalent bond contains a shared pair of electrons (see Fig 9).

Multiple covalent bonds contain multiple pairs of electrons:

- a double covalent bond contains 2 shared electron pairs (see Fig 9),
- a triple covalent bond contains 3 shared electron pairs.

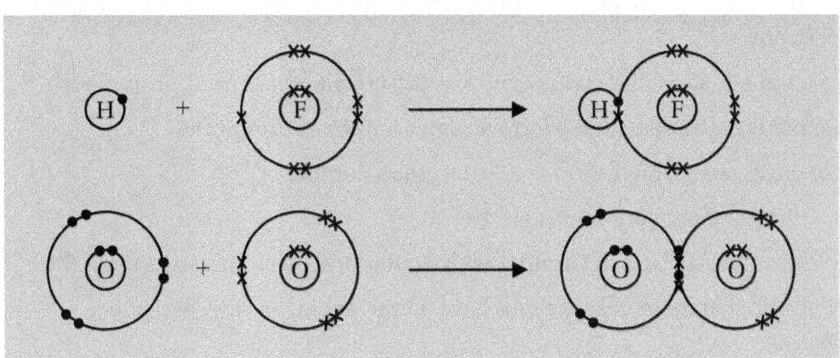

Fig 9
Formation of a single covalent bond (hydrogen-fluorine)
Formation of a double covalent bond (oxygen-oxygen)

In each molecule, the electron pairs create 'bonds' between the two atoms because they attract the nucleus of each atom and therefore resist the separation of the two atoms.

A shared electron pair in a single covalent bond is usually represented like this: H—F.

A double covalent bond is usually represented like this: O=O.

A triple covalent bond is usually represented like this: N≡N.

A **co-ordinate bond** (or **dative covalent bond**) contains a shared pair of electrons, with both electrons supplied by one atom, using a non-bonding electron pair (lone pair). As soon as it is formed, a dative covalent bond is identical to a normal covalent bond (see Fig 10).

Co-ordinate bonds are usually represented by an arrow, thus →.

The direction of the arrow shows the direction of movement of the electron pair to form the dative covalent bond.

For example, the fourth N to H bond in the ammonium ion can be shown like this: $[H_3N{\rightarrow}H]^+$.

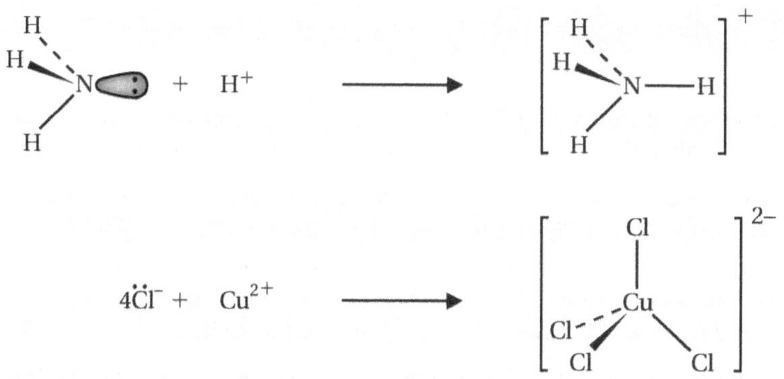

H_3N: + H^+ → $[H_3N—H]^+$

$4\ddot{Cl}^- + Cu^{2+}$ → $[CuCl_4]^{2-}$

Fig 10
Co-ordinate bond formation

Essential Notes

Note that both of these ions have a tetrahedral shape, but are viewed from different angles.

The wedge represents a covalent bond coming out in front of the plane of the paper, and the dashed line a covalent bond going behind the plane of the paper.

(lone pair symbol) represents a lone, non-bonding pair of electrons.

3.1.3.3 Metallic bonding

Metallic bonds in solids do not exist in isolation. They form part of a **giant metallic lattice** which consists of close-packed metal ions surrounded by **delocalised electrons**, which are free to move through the lattice. Fig 11 shows a metallic lattice in two dimensions.

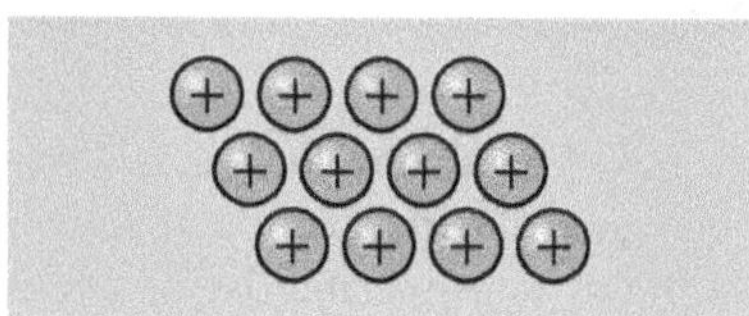

Fig 11
A slice through a three-dimensional metallic lattice. The positive metal ions (cations) are held together by delocalised electrons (not shown)

Metals usually have high melting and boiling points, because a large amount of energy must be supplied in order to remove a metal atom from the attraction of the delocalised electrons.

Essential Notes

Delocalised electrons occupy an orbital which spreads across all the atoms that make up a metal crystal.

3.1.3.4 Bonding and physical properties

Types of crystal

A solid with a regular shape which contains particles organised in a regular structure is called a crystal. Crystals can be classified according to the type of bonding between particles.

Ionic crystals. In the ionic lattice formed by sodium chloride, each sodium ion has six chloride ions as nearest neighbours and each chloride ion is surrounded by six sodium ions (see Fig 12).

Ionic crystals usually have high melting points because the ions are held in position by strong electrostatic forces.

Fig 12
The sodium chloride lattice

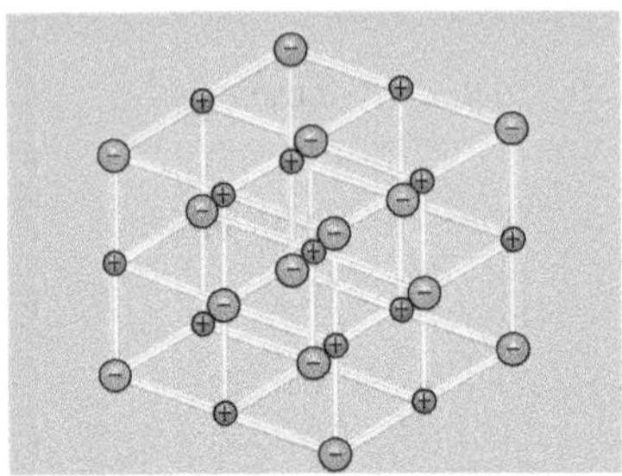

Ionic crystals are hard and brittle. They do not conduct electricity when solid, but when molten the ions are free to move and can carry a current.

Metallic crystals. In these crystals, the ions are usually packed together as closely as possible (see Fig 13). This means that each ion has six nearest neighbours in the same plane, three above and three below, making twelve nearest neighbours in total. The metallic bonds between ions are usually quite strong and most metals have high melting and boiling points.

Metals are malleable (can be hammered into shape) and ductile (can be drawn into wires) because the planes of ions can slide over each other.

Fig 13
Metallic structure of magnesium: a two-dimensional representation, showing the doubly charged metal ions (cations) held together by delocalised electrons (not shown)

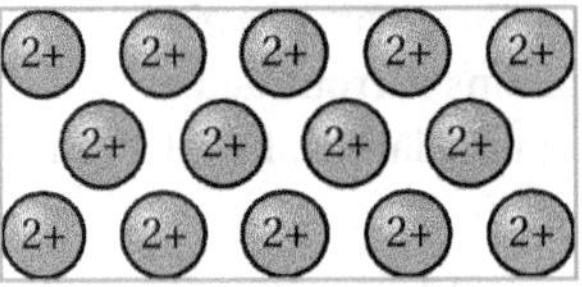

The bonding electrons are free to move between ions (the electrons are delocalised) leading to good electrical conductivity in the solid state. The delocalised electrons are also responsible for the ability of a metal to reflect light (metallic lustre).

Macromolecular (giant covalent) crystals. The diamond crystal is very hard and has a high melting point, because all the carbon atoms are linked together by strong covalent bonds to form a giant crystal or **macromolecule** (see Fig 14).

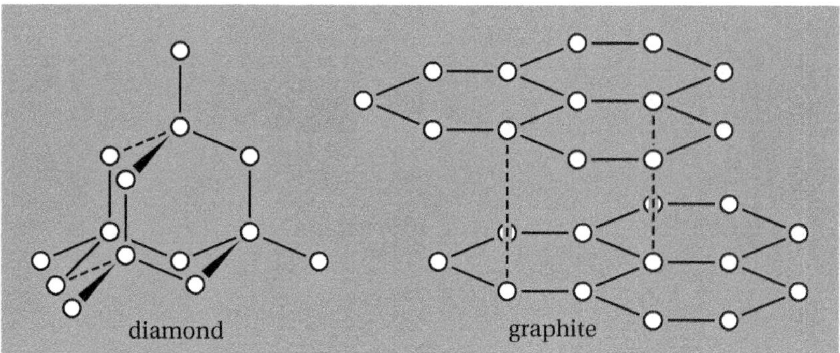

Fig 14
Structures of the macromolecules diamond and graphite

There are no covalent bonds between the layers in graphite. The dotted lines have been drawn to show the position of the second layer of carbon atoms relative to the first.

Graphite has a high melting point because all the atoms in each plane are linked by strong covalent bonds. The forces between the planes are weak, so the planes can slide over each other and, unlike most other macromolecules, graphite crystals are therefore soft. Exceptionally, for a macromolecule, graphite is able to conduct electricity because there are delocalised electrons above and below each plane of atoms. The delocalised electrons are free to move parallel to the planes.

Molecular crystals. The covalent molecules in molecular crystals are held together by one or more of the following interactions: weak van der Waals' forces, dipole-dipole forces or hydrogen bonding. In an iodine crystal (Fig 15) the molecules are arranged in a regular array, but the forces between the molecules are weak van der Waals' forces so that the crystal has a low melting point.

Ice also has the molecules arranged in a regular array (see Fig 16), but the forces of attraction between molecules are stronger hydrogen bonds, so the crystal has a higher melting point than iodine.

Properties of solids

The different types of crystal can be recognised by their physical properties as summarised in Table 5.

Nature of gases, liquids and solids

Gases are made up of particles that are quite far apart and which move with rapid random motion. The size of the particles and any intermolecular forces can be ignored unless the particles are close together, at high pressure or at low temperature.

Notes

Macromolecules melt at very high temperatures. Graphite is unique; most macromolecules are non-conductors under all conditions. However, some electrically-conducting polyethynes have been synthesised and used in rechargeable batteries and in electroluminescent devices.

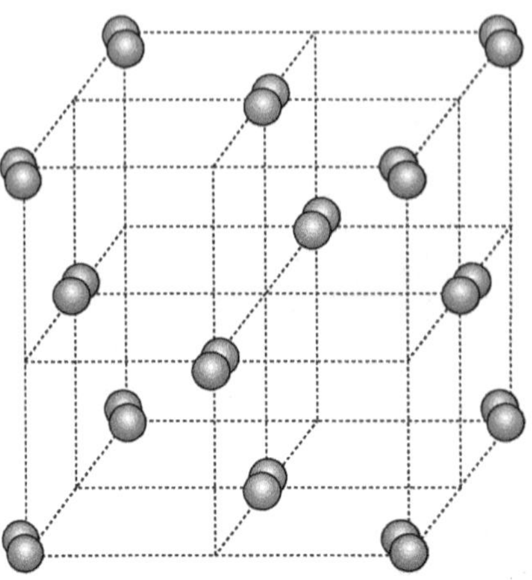

Fig 15
Molecular lattice of iodine

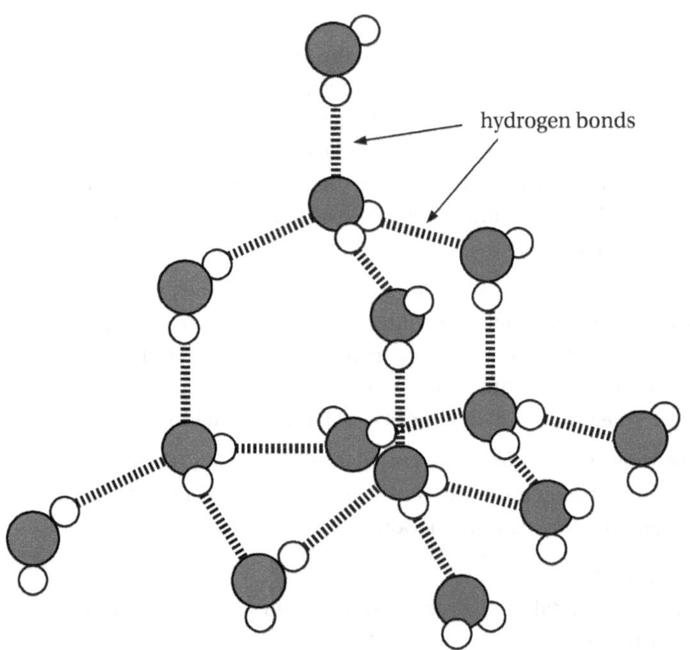

Fig 16
Molecular lattice of ice

Table 5
Some physical properties of crystals

Type of crystal	*T_m and T_b (relative values)*	*Electrical conductivity when solid*	*Electrical conductivity when molten*	*Solubility in water*
ionic	high	non-conductor	good	variable but often good
macromolecular	very high	non-conductor (except graphite)	non-conductor	insoluble
molecular	low	non-conductor	non-conductor	variable
metallic	usually high	good	good	insoluble

Essential Notes

T_m is the melting point or melting temperature.

Essential Notes

T_b is the boiling point or boiling temperature.

In liquids, the particles are in a state of order intermediate between that of a gas and that of a solid. At any instant in time the arrangement of particles resembles a somewhat disordered solid. Over a period of time the disordered regions allow all the particles in the liquid to move through the liquid. The particles are held together by forces similar to those in a solid.

In solids, the particles remain in fixed positions, about which they can vibrate. The forces which hold the particles together can be ionic attractions, covalent bonds, metallic bonds, hydrogen bonds, dipole-dipole forces or van der Waals' forces.

Heat energy is required to change a solid into a liquid at its melting point. The energy is used to *loosen* the forces which hold the particles together. This heat energy is called the **enthalpy of fusion**.

More energy is needed to change *phase* from a liquid into a gas than to change from a solid into a liquid. The energy is used to *overcome* the forces which hold the particles together so that the particles can be completely separated. This heat energy is called the **enthalpy of vaporisation**.

Note: the precise nature of the intermolecular forces referred to in this section is considered in section 3.1.3.7.

Essential Notes

The separation of particles in liquids is usually only about 10% more than in solids. In gases, separations are much larger.

Essential Notes

The term *phase* is most commonly used to describe transitions between solid, liquid and gasous states of matter.

3.1.3.5 Shapes of simple molecules and ions

The outer electrons of atoms in molecules are arranged in pairs. These electron pairs can be considered as 'clouds' of electron density which repel each other, so that they are as far apart as possible. Such repulsions lead to the arrangements of electron pairs and the bond angles shown in Table 6.

Table 6
Shapes of molecules

Number of pairs of electrons	*Arrangement*	*Angles*	*Name of shape*
2 pairs	X	180°	linear
3 pairs	X	120°	trigonal planar
4 pairs	X	109.5°	tetrahedral
5 pairs	X	90°, 120°	trigonal bipyramidal
6 pairs	X	90°	octahedral

Notes

It is the repulsion between electron pairs which dictates the shape of a molecule or ion, **not** the repulsion between atoms.

The shapes of molecules and ions which contain only *single covalent bonds* between their atoms can therefore be predicted from the total number of electron pairs in the outside shell of the central atom. This number of electron pairs is calculated by taking into consideration:

- the number of outside-shell electrons originally in the central atom
- the number of additional shared electrons in covalent or co-ordinate bonds
- the loss or gain of additional electrons if the species is a positive or a negative ion.

Essential Notes

Lone pair–lone pair repulsions are stronger than lone pair–bonding pair repulsions. These in turn are stronger than bonding pair–bonding pair repulsions.

The final shape is also modified if some of the electron pairs are lone (non-bonding) pairs. Lone pairs are more compact than bonding pairs so they repel each other and other pairs more strongly, leading to bond angles between bonding pairs which are *smaller* than those found in totally symmetrical shapes. These principles can be used to predict different shapes, as illustrated in Table 7.

Table 7
Determination of the shapes of molecules and ions

Molecule or ion	*Outside shell electrons*	*Total number of electrons*	*Number of electron pairs*	*Shape*
BF_3	3 from B + 1 from each F	6	3	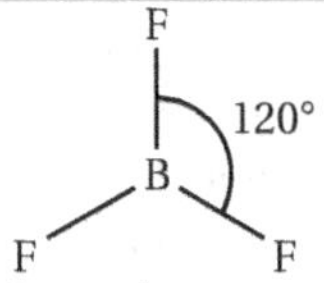 F, B, F, F, 120°; the three electron pairs repel equally
CH_4	4 from C + 1 from each H	8	4	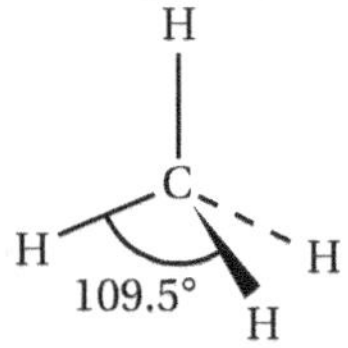 H, C, H, H, H, 109.5°
NH_3	5 from N + 1 from each H	8	4	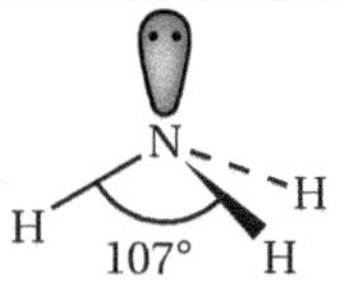 N, H, H, H, 107°; the single lone pair does not repel as strongly as the 2 lone pairs in H_2O
H_2O	6 from O + 1 from each H	8	4	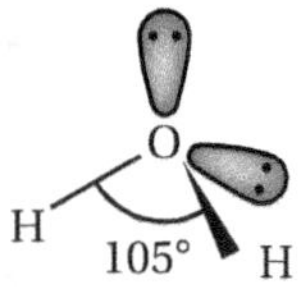 O, H, H, 105°; the lone pairs repel more strongly than the bonding pairs
PF_5	5 from P + 1 from each F	10	5	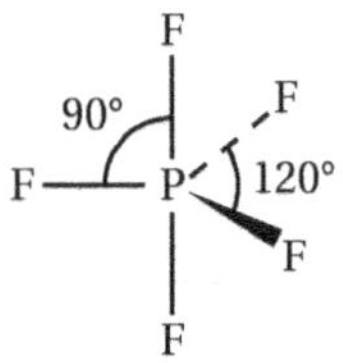 F, F, F, P, F, F, 90°, 120°

Essential Notes

The bond angle in water is less than the tetrahedral angle.

Molecule or ion	Outside shell electrons	Total number of electrons	Number of electron pairs	Shape
SF_6	6 from S + 1 from each F	12	6	
ClF_4^-	7 from Cl + 1 from each F + 1 negative ion	12	6	the lone pairs repel most, therefore they are as far apart as possible

Essential Notes

In this ion the electron pairs are arranged octahedrally. The shape of the ion is described as **square planar**.

3.1.3.6 Bond polarity

Electronegativity

Electronegativity is the power of an atom to attract the electrons in a covalent bond. It can be calculated by various means and is usually given a number ranging from 0.7 to 4.0. Small atoms with a large number of protons in the nucleus attract bonding electrons most strongly. Therefore electronegativity increases from left to right across a period in the Periodic Table, and from the bottom to the top of a group (Table 8).

Essential Notes

In the Periodic Table electronegativity increases in these directions

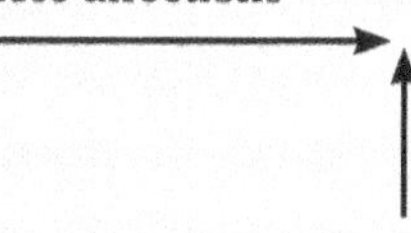

H 2.1							He
Li 1.0	Be 1.5	B 2.0	C 2.5	N 3.0	O 3.5	F 4.0	Ne
Na 0.9	Mg 1.2	Al 1.5	Si 1.8	P 2.1	S 2.5	Cl 3.0	Ar
						Br 2.8	Kr

Table 8
Electronegativity values

Polar covalent bonds

When a covalent bond exists between atoms of differing electronegativity, the shared pair of electrons is displaced towards the more electronegative atom:

$\delta+$ $\delta-$
A—B **B** is more electronegative than **A**

The displacement of electron density makes the less electronegative atom slightly electron-deficient (hence $\delta+$), while the more electronegative atom has a slight excess of electron density (hence $\delta-$). This charge separation creates an electric 'dipole' and the bond is described as **polar**. Three examples of molecules with polar bonds are shown in Fig 17.

Notes

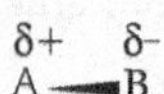

This diagram shows how the electron pair in a polar bond is displaced towards the more electronegative atom.

Fig 17
Polar and non-polar molecules

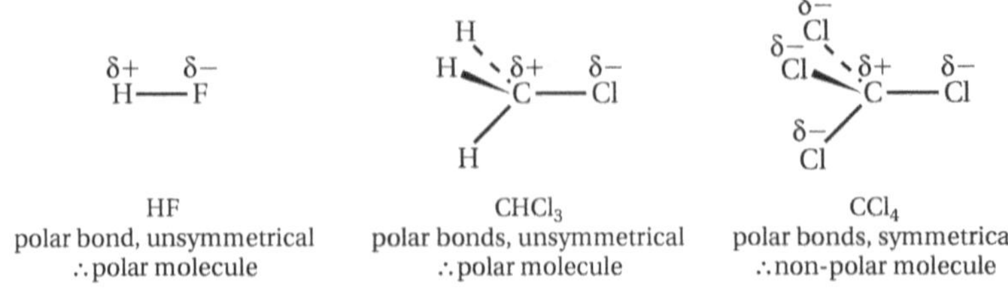

Notes

Note that it is **incorrect** to represent the polar hydrogen chloride molecule with full charges like this:

$\overset{+}{H}—\overset{-}{Cl}$

The polarity of a bond can be measured in a unit called the debye. Its magnitude depends on the difference in electronegativity between elements (shown in Table 9).

Table 9
Polar molecules

Molecule	Electronegativity difference	Dipole/debye
HCl	0.9	1.03
HBr	0.7	0.78
HI	0.4	0.38

Unsymmetrical molecules with polar bonds, like those shown in Fig 17, have a permanent dipole. That is, one side of the molecule is permanently slightly positive and the other side permanently equally negative.

Symmetrical molecules, like CCl_4 (which is tetrahedral) contain polar bonds but are non-polar molecules. They have no permanent dipole as the polarities cancel each other out.

3.1.3.7 Forces between molecules

Covalent molecules are attracted to each other by intermolecular forces. The three types of intermolecular force are:

induced dipole–dipole (van der Waals' forces, dispersion forces, London forces)	permanent dipole–dipole	hydrogen bonding

weakest ⟶ strongest

This order of strength is true only for small molecules.

All species, even noble gas atoms, are attracted to each other by **van der Waals' forces** (also known as **dispersion forces** or **London forces**).

Polar molecules contain atoms with different electronegativities and, in addition to attraction by van der Waals' forces, these molecules attract each other by **permanent dipole–dipole forces**.

Molecules which contain hydrogen covalently bonded to a nitrogen, oxygen or fluorine atom are attracted to each other by hydrogen bonding. They are also attracted to each other by permanent dipole–dipole forces and van der Waal's forces.

Essential Notes

In larger molecules, van der Waals' forces are usually stronger than permanent dipole-dipole interactions.

Notes

For similar molecules, the magnitudes of the van der Waals' forces increase with relative molecular mass.

Van der Waals' forces

Van der Waals' forces are *temporary dipole–induced dipole attractions.* At any instant in time, the electron distribution in a non-polar covalent molecule may be asymmetrical, owing to the fluctuating movement of electrons. This leads to a temporary dipole which induces an opposite dipole on an adjacent molecule. The second molecule is therefore attracted to the first molecule. The magnitude of van der Waals' forces increases with the size of molecules and also depends upon their shape. Branched-chain hydrocarbons have weaker intermolecular forces than straight-chain molecules because they are less polarisable.

Fig 18 helps to illustrate these effects. The ellipses represent the boundaries of the electrons in the molecules. The δ+ and δ− charges shown are temporary and fluctuate around the molecules with time.

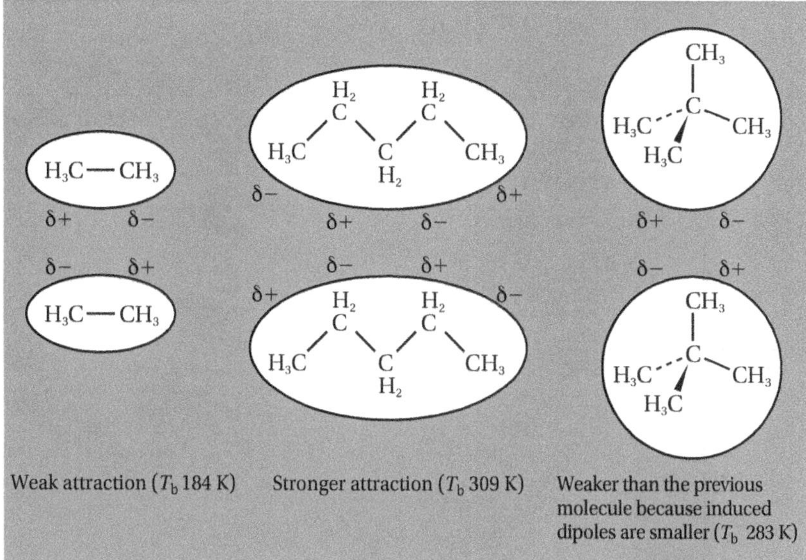

Fig 18
Van der Waals' forces in alkanes

Notes

2,2-Dimethylpropane and pentane have the same formula (C_5H_{12}) and the same relative molecular mass, but the molecules of the branched-chain alkane, owing to their more spherical shape, cannot pack together as closely and therefore their induced dipoles are weaker.

Permanent dipole-dipole forces

Molecules with permanent dipoles attract each other as shown in Fig 19.

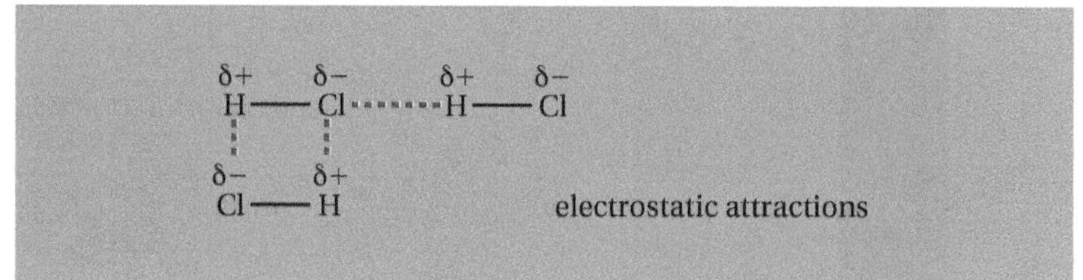

Fig 19
Dipole-dipole attractions

Hydrogen bonding

Hydrogen bonding is the name given to the strongest type of intermolecular force between neutral molecules. It is a special case of a dipole-dipole force that exists between a lone pair of electrons on a N, O or F atom and a hydrogen atom that has a strong partial charge (δ+), because it is attached to an atom with a large electronegativity (N, O or F). The electronegative atom pulls electrons away from the hydrogen so that, on the opposite side to the bond, the hydrogen appears almost like an unshielded proton. Two examples of hydrogen bonding are shown in Fig 20.

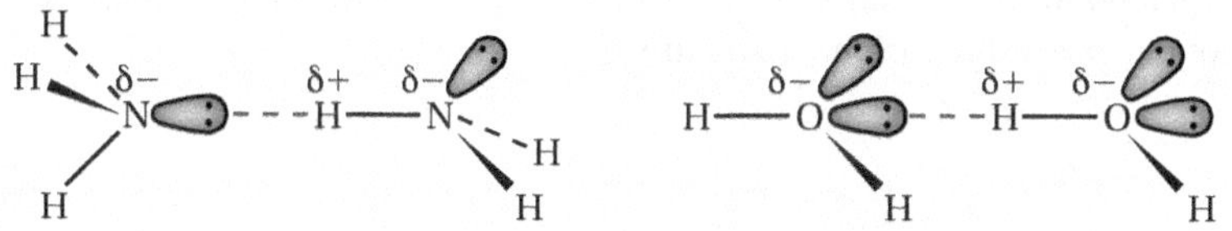

Fig 20
Hydrogen bonding

Other examples of molecules which hydrogen-bond to each other are HF, CH_3CH_2OH and CH_3COOH. These substances all have relatively high boiling points due to the hydrogen bonding.

Notes

The nucleus of a hydrogen-bonded hydrogen atom is always in line with the nuclei of the two electronegative atoms on either side.

Hydrogen bonds are much weaker than covalent bonds – typically between 5% and 10% of the strength of a covalent bond. Nevertheless, this intermolecular force is strong enough to cause unusually high boiling points for some compounds (Fig 21).

Fig 21
Boiling points of Group 6 and Group 7 hydrides

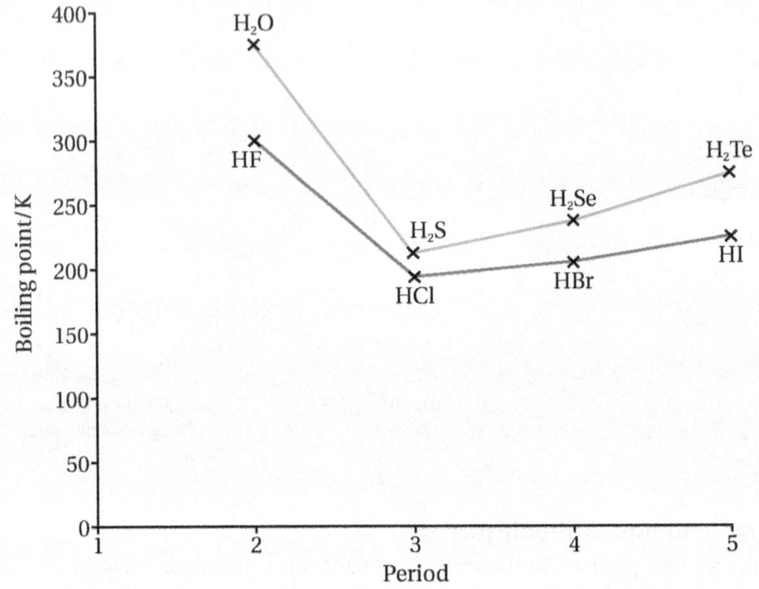

The boiling points of hydrides generally increase down a group in the Periodic Table. The boiling points of HF and H_2O go against this trend, being higher than expected owing to hydrogen bonding.

Hydrogen bonding is also important because it influences the structures of some solids, such as the structure of ice and the shapes of some proteins.

In ice, hydrogen bonding holds the water molecules together in a three-dimensional array that occupies more space than in liquid water. This explains why ice is less dense than water and therefore floats.

Fig 22 is a simplified representation of the arrangement of water molecules in ice. The non-bonding electron pairs on the oxygen atoms are not shown and one of the hydrogen bonds to each oxygen has been omitted. In ice, the ring of six water molecules is not planar.

A more complete representation would show a three-dimensional structure, with each oxygen linked to four hydrogens using two covalent and two hydrogen bonds in a tetrahedral arrangement. Fig 23 shows in more detail a part of this three-dimensional structure.

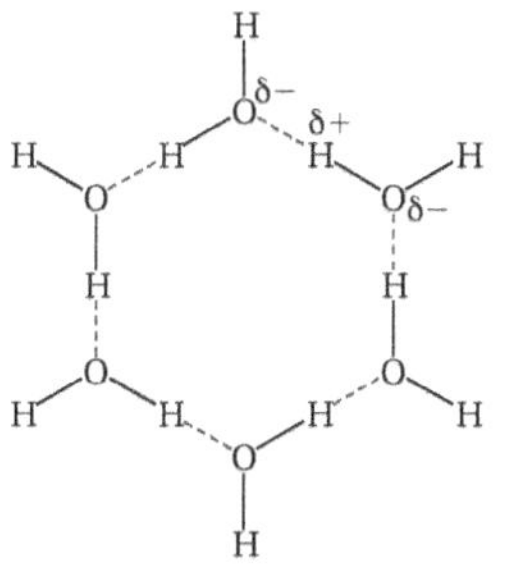

Fig 22
A simplified representation of six water molecules in an ice crystal.

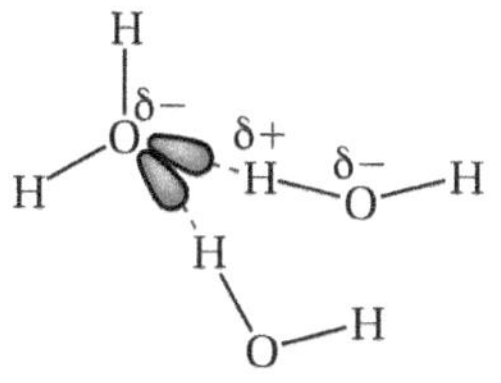

Fig 23
The three-dimensional nature of hydrogen bonding in ice or water (simplified)

Hydrogen bonding is also responsible for some of the forces that hold protein molecules together and for protein molecules adopting the arrangement of a helix or a *'pleated sheet'* (see Fig 24).

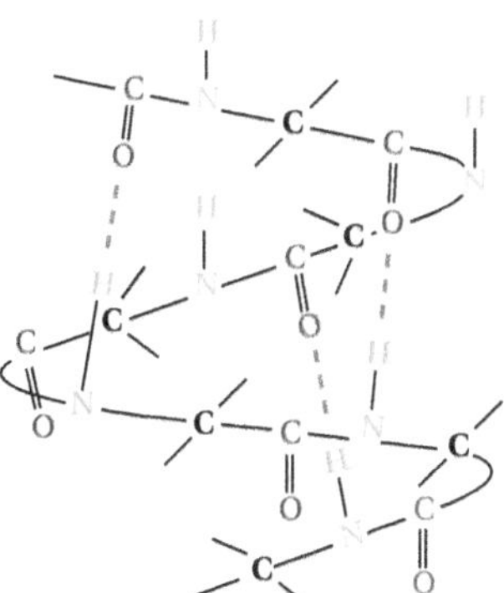

Fig 24
Hydrogen bonding in protein molecules (simplified)

3.1.4 Energetics

3.1.4.1 Enthalpy change

The **enthalpy change** of a system is the heat energy change at constant pressure. It is indicated by the symbol ΔH where Δ is the 'change in' or 'difference in' and H is the enthalpy.

When a system gives out heat energy to the surroundings, enthalpy is lost by the system so that ΔH is negative (**exothermic**).

When a system takes in heat from the surroundings, enthalpy is gained by the system so that ΔH is positive (**endothermic**).

Standard enthalpy changes occur at the standard pressure of 100 kPa (1 bar) and a stated temperature, usually taken as 298 K. The symbol $\Delta H^{\ominus}$ is used for standard enthalpy changes.

An element or a compound is said to be in its **standard state** when it is in its normal, stable state at 298 K and 100 kPa. Where a compound such as water could be either a gas or a liquid under standard conditions, the physical state of the substance should be clarified by symbol or by explanation.

For example, $H_2O(l)$ refers to water in the liquid state, $H_2O(g)$ refers to water vapour (steam). Where elements exist in allotropic forms, the particular **allotrope** should be specified, for example C (graphite) or C (diamond). If the allotrope is not specified it is assumed to be the more stable form (graphite in the case of carbon).

Notes

The sign of the enthalpy change is taken from the point of view of the reaction. If the reaction gives out (loses) heat energy, the enthalpy change is negative. If the reaction takes in (gains) heat energy, the enthalpy change is positive.

Standard enthalpy of combustion $\Delta_c H^\circ$

Definition

*The **standard enthalpy of combustion** is defined as the enthalpy change, under standard conditions, when 1 mol of a substance is burned completely in oxygen, with all reactants and products in their standard states.*

Essential Notes

The symbol ⦵ shows that the change is measured under standard conditions.

Standard conditions are usually taken as 100 kPa and 298 K; at this temperature, water is usually taken to be a liquid.

The enthalpy change is linked to an equation with state symbols. For example:

$$CH_4(g) + 2O_2(g) \rightarrow CO_2(g) + 2H_2O(l) \qquad \Delta_c H^\circ = -890 \text{ kJ mol}^{-1}$$

Enthalpies of combustion are determined experimentally using a **calorimeter**.

Notes

The measurement of an enthalpy change is a required practical activity.

Standard enthalpy of formation $\Delta_f H^\circ$

Definition

*The **standard enthalpy of formation** is defined as the enthalpy change, under standard conditions, when 1 mol of a compound is formed from its elements, with all reactants and products in their standard states.*

Essential Notes

Accurate enthalpies of combustion are determined by an experiment in a bomb calorimeter. In a school laboratory it is possible to measure enthalpies of combustion using simple apparatus such as a 'spirit burner' but this usually gives values which are not sufficiently exothermic. The main error is due to an inability to measure the heat energy which is lost to the surroundings.

By definition, for an element the standard enthalpy of formation must be zero.

The following is an example of a reaction for which the enthalpy change is the enthalpy of formation:

$$2Na(s) + C(\text{graphite}) + \tfrac{3}{2}O_2(g) \rightarrow Na_2CO_3(s) \quad \Delta_f H^\circ = -1131 \text{ kJ mol}^{-1}$$

Enthalpies of formation are usually determined indirectly using Hess's law, as explained later, in section 3.1.4.3, and can be found in data-book tables.

3.1.4.2 Calorimetry

The heat energy, q, required to change the temperature of a substance by an amount ΔT can be calculated using the expression:

$$q = m \times c \times \Delta T$$

where m is the mass of the substance and c is the specific heat capacity. Commonly, c is given with the units kJ K^{-1} kg^{-1} which requires that m be expressed in kilograms and ΔT in kelvin, giving q the units kJ. For many chemical reactions in aqueous solution it can be assumed that the only substance heated is water, which has a specific heat capacity of 4.18 kJ K^{-1} kg^{-1}. The value of the specific heat capacity of a substance, c, will always be given for use in calculations.

The heat energy, q, can be used to calculate an enthalpy change as shown in the two examples that follow.

Example

In an experiment, 1.00 g of methanol (CH_3OH) was burned in air and the flame was used to heat 100 g of water, which rose in temperature by 42.0 °C.

$$CH_3OH(l) + \frac{3}{2}O_2(g) \rightarrow CO_2(g) + 2H_2O(g)$$

Calculate the enthalpy change.

Answer

- Heat energy gained by the water $q = m \times c \times \Delta T$
  $= 0.100 \times 4.18 \times 42.0$
  $= 17.6$ kJ
- Heat energy lost by methanol $= -17.6$ kJ
- Moles of methanol burned $= \dfrac{\text{mass}}{M_r}$
  $= \dfrac{1.00}{32.0} = 0.0313$ mol
- Enthalpy change per mole $\Delta H = \dfrac{\text{heat energy lost by methanol}}{\text{moles of methanol}}$
  $= \dfrac{-17.6}{0.0313}$ kJ mol^{-1}
  $= -563$ kJ mol^{-1}

Essential Notes

For the purpose of this calculation, heat losses are ignored and the heat absorbed by the water container is regarded as negligible.

Notes

Note that the mass of water must be converted into kg (100 g = 0.100 kg) but that a temperature difference in degrees Celsius is the same as that in kelvin.

Notes

Note the *negative* sign because the reaction is exothermic.

Example

In an insulated container, 50 cm^3 of 2.00 mol dm^{-3} HCl at 293 K were added to 50.0 cm^3 of 2.00 mol dm^{-3} NaOH also at 293 K. After reaction, the temperature of the mixture rose to 307 K.

$$HCl(aq) + NaOH(aq) \rightarrow NaCl(aq) + H_2O(l)$$

Calculate the enthalpy change.

Answer

- Temperature rise $\Delta T = 14.0$ K
- Heat energy gained by the water $q = m \times c \times \Delta T$
  $= 0.100 \times 4.18 \times 14.0$
  $= 5.85$ kJ
- Heat energy lost by the reaction $= -5.85$ kJ
- Moles of acid $=$ volume (in dm^3) $\times$ concentration
  (= moles of alkali) $= \dfrac{50.0 \times 2.00}{1000} = 0.100$ mol

Notes

The total volume of water in the reaction mixture is 100 cm^3. This has a mass of 0.10 kg. The amount of water produced by the reaction is negligibly small. The heat capacity of the solution is assumed to be the same as that of water.

Notes

The enthalpy change is usually related to the 'moles of equation' as written. Again, this is an exothermic reaction so the sign of the enthalpy change is negative.

- Enthalpy change $\Delta H = \dfrac{\text{heat energy lost by the reaction}}{\text{moles of acid}}$

$$= \frac{-5.85}{0.100} \text{ kJ mol}^{-1}$$

$$= -58.5 \text{ kJ mol}^{-1}$$

3.1.4.3 Applications of Hess's law

Essential Notes

The first law of thermodynamics is also similar to the principle of conservation of energy.

Definition

***The first law of thermodynamics** states that energy can be neither created nor destroyed, but can be converted from one form into another.*

Hess's law is a special case of the first law.

Essential Notes

The overall enthalpy change for a multi-step reaction can be calculated using the expression

$$\Delta H = (\Delta H_{(\text{step 1})} + \Delta H_{(\text{step 2})} + \ldots) = \Sigma \Delta H_{(\text{all steps})}$$

where the symbol Σ means 'sum of'.

Definition

***Hess's law** states that the enthalpy change of a reaction depends only on the initial and final states of the reaction and is independent of the route by which the reaction occurs.*

It follows from Hess's law that the enthalpy change of a reaction is the sum of the individual enthalpy changes of each step into which the reaction can be divided, regardless of their nature.

Hess's law is illustrated diagrammatically in Fig 25.

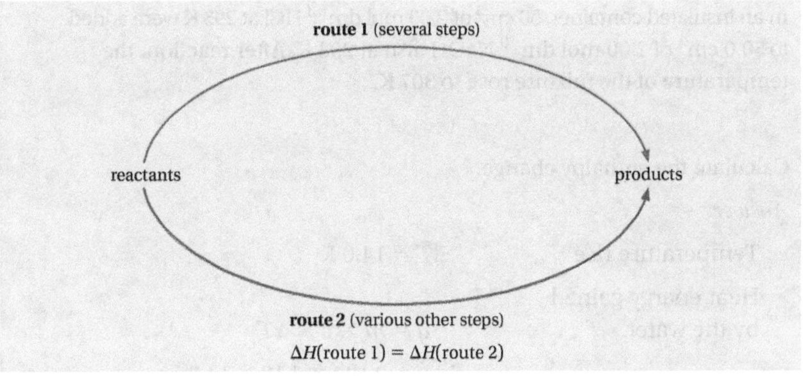

Fig 25
Hess's law in terms of a heat cycle

Hess's law can be used to determine ΔH values for reactions where direct determination is difficult. For example, the enthalpy change for any reaction can be determined if the enthalpies of combustion of the reactants and the products are known. Thus, the standard enthalpy of formation of methane can be calculated from standard enthalpies of combustion as shown in Fig 26.

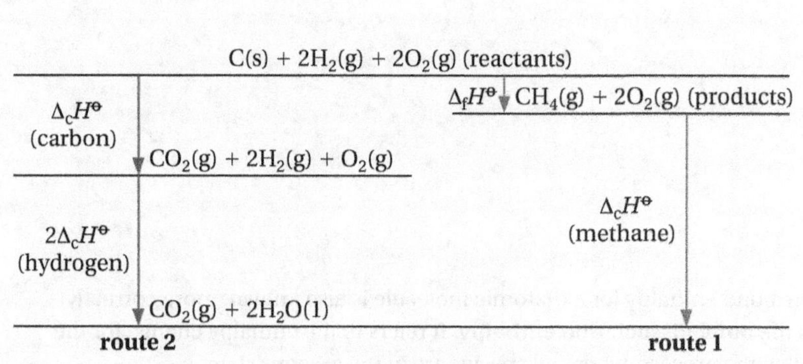

Fig 26
Using an enthalpy cycle to determine the enthalpy of formation of methane

Essential Notes

In the step where CH_4 is formed from C(s) and $2H_2(g)$, the oxygen can be ignored because it is present in both the reactants and products.

Notes

Remember $\Delta_f H^\ominus(O_2)$ is zero because oxygen is an element.

This cycle shows how an enthalpy change can be calculated from standard enthalpies of combustion. Note that in these diagrams, unlike the way equations are usually written, reactants are on one horizontal line and the products are on another.

$$\Sigma \Delta H(\text{steps in route 1}) = \Sigma \Delta H(\text{steps in route 2})$$

$$\therefore \Delta_f H^\ominus(\text{products}) + \Delta_c H^\ominus(\text{products}) = \Sigma_c H^\ominus(\text{reactants})$$

$$\therefore \Delta_f H^\ominus(\text{products}) = \Sigma \Delta_c H^\ominus(\text{reactants}) - \Sigma_c H^\ominus(\text{products})$$

$$\therefore \Delta_f H^\ominus(\text{methane}) = \Delta_c H^\ominus(\text{carbon}) + 2 \times \Delta_c H^\ominus(\text{hydrogen}) - \Delta_c H^\ominus(\text{methane})$$

$$= -393 + (2 \times -285) - (-890)$$

$$= -73 \text{ kJ mol}^{-1}$$

Notes

In general, for any reaction
$\Delta H^\ominus = \Sigma \Delta_c H^\ominus$ (reactants) – $\Sigma \Delta_c H^\ominus$ (products)

Enthalpy changes for reactions can also be determined from tabulated values of enthalpies of formation. For example, the enthalpy change for the reaction:

$$3CO(g) + Fe_2O_3(s) \rightarrow 2Fe(s) + 3CO_2(g)$$

can be determined as shown in Fig 27.

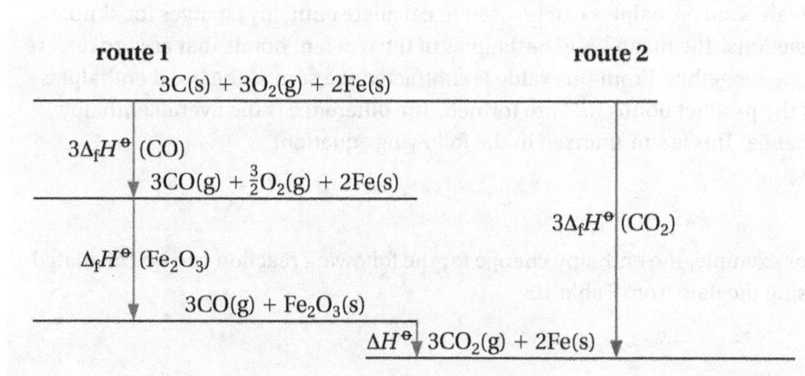

Fig 27
Using an enthalpy cycle to determine the enthalpy change for a reaction

Notes

This cycle shows how an enthalpy change can be calculated from standard enthalpies of formation.

Note that $\Delta_f H^\ominus$(Fe) is zero.

Notes
In general for any reaction
$\Delta H^{\ominus} = \Sigma\Delta_f H^{\ominus}(\text{products}) - \Sigma\Delta_f H^{\ominus}(\text{reactants})$
Note that this expression is different from the one which involves enthalpies of combustion.

$\Sigma\Delta H(\text{route 1}) = \Sigma\Delta H(\text{route 2})$

$\therefore \Delta H^{\circ} + \Sigma\Delta_f H^{\circ}(\text{reactants}) = \Sigma\Delta_f H^{\circ}(\text{products})$

$\therefore \Delta H^{\circ} = \Sigma\Delta_f H^{\circ}(\text{products}) - \Sigma\Delta_f H^{\circ}(\text{reactants})$

$\therefore \Delta H^{\circ} = 3 \times \Delta_f H^{\circ}(CO_2) - (3 \times \Delta_f H^{\circ}(CO) + \Delta_f H^{\circ}(Fe_2O_3))$

$= 3 \times -394 - ((3 \times -111) - 822)$

$= -27\ \text{kJ mol}^{-1}$

3.1.4.4 Bond enthalpies

The bond enthalpy for a diatomic molecule is also known, more correctly, as the **bond dissociation enthalpy**. It refers to the enthalpy change for the following process, where all species are in the gaseous state.

$A{-}B(g) \rightarrow A(g) + B(g) \qquad \Delta H = \text{bond enthalpy}$

In polyatomic molecules it is convenient to use the term **mean bond enthalpy**.

Definition
***The mean bond enthalpy** is the average of several values of the bond dissociation enthalpy for a given type of bond, taken from a range of different compounds.*

Essential Notes
Bond enthalpy calculations apply only to reactions in the gaseous state.

Consider the following processes:

$CH_4(g) \rightarrow CH_3(g) + H(g) \qquad \Delta H = 423\ \text{kJ mol}^{-1}$

$CH_4(g) \rightarrow C(g) + 4H(g) \qquad \Delta H = 1664\ \text{kJ mol}^{-1}$

Notes
The mean bond enthalpy per C—H bond is $\frac{1664}{4}$ = 416 kJ mol^{-1}.

In data books, the mean bond enthalpy refers to the average of the bond enthalpy values for many different compounds.

The second equation involves the breaking of all four carbon-hydrogen bonds. The mean bond enthalpy can therefore be determined by dividing the value of 1664 by 4. The bond dissociation enthalpy value for breaking the H—CH_3 bond (423 kJ mol^{-1}) is slightly different from the value for the mean bond enthalpy (416 kJ mol^{-1}). The first process involves the breaking of one C—H bond and the formation of a •CH_3 radical. The second process does not involve the formation of hydrocarbon radicals; it leads only to atomic species. The mean bond enthalpy is a useful quantity when calculating reaction enthalpy changes, but its use is only approximate.

Mean bond enthalpies can be used to calculate enthalpy changes for simple reactions. The mean bond enthalpies of the reactant bonds that are broken are added together. From this value is subtracted the sum of the bond enthalpies of the product bonds that are formed. The difference is the overall enthalpy change. This is summarised in the following equation:

$$\Delta H = \Sigma(\text{mean bond enthalpy of bonds broken}) - \Sigma(\text{mean bond enthalpy of bonds formed})$$

Notes
This calculation assumes that the C—H bond enthalpy in CH_3Cl is equal to that in CH_4; this is a good approximation.

For example, the enthalpy change for the following reaction can be calculated using the data from Table 10:

$CH_4(g) + Cl_2(g) \rightarrow CH_3Cl(g) + HCl(g)$

$\Delta H = \Sigma(\text{enthalpy of bonds broken}) - \Sigma(\text{enthalpy of bonds formed})$

$$\Delta H = (\text{C—H} + \text{Cl—Cl}) - (\text{C—Cl} + \text{H—Cl})$$
$$= (412 + 242) - (338 + 431)$$
$$= -115 \text{ kJ mol}^{-1}$$

Alternatively, the enthalpy change for this reaction can also be calculated from other thermochemical data, for example, the $\Delta_f H^\ominus$ values of the compounds:

As seen at the end of section 3.1.4.3,

$$\Delta H^\ominus = \Sigma\Delta_f H^\ominus(\text{products}) - \Sigma\Delta_f H^\ominus(\text{reactants})$$
$$\Delta H^\ominus = \Delta_f H^\ominus(CH_3Cl) + \Delta_f H^\ominus(HCl) - \Delta_f H^\ominus(CH_4) - \Delta_f H^\ominus(Cl_2)$$
$$\Delta H^\ominus = (-84) + (-92) - (-75) - (0)$$
$$\Delta H^\ominus = -101 \text{ kJ mol}^{-1}$$

It can be seen that there is a small difference between the value of ΔH obtained from mean bond enthalpies and the value of ΔH obtained from thermochemical data (in this case the $\Delta_f H^\ominus$ values).

Of the two values, that derived from thermochemical data is considered to be the more reliable. The thermochemical data are obtained directly from the actual compounds involved in the reaction, whereas bond enthalpy values are average values calculated from the measured bond enthalpies of a range of similar compounds.

Typically, as here, the difference between the ΔH values is only a small one, but when a ΔH value based on mean bond enthalpy data is significantly different to that calculated from direct thermochemical data, then assumptions made about the precise nature of the bonds involved in one or more of the substances may be brought into question.

Notes

It is not always necessary to consider all the bonds in the reactants and products. In this example, the answer can be determined by considering only the bonds broken and those formed.

Essential Notes

Remember that, by definition, $\Delta_f H^\ominus$ of an element is always 0.

Bond	C—H	C—Cl	Cl—Cl	H—Cl
Mean bond enthalpy/kJ mol^{-1}	412	338	242	431

Table 10
Mean bond enthalpies

3.1.6 Chemical equilibria, Le Chatelier's principle and K_c

3.1.6.1 Chemical equilibria and Le Chatelier's principle

Dynamic nature of equilibrium

Many chemical reactions continue until one of the reactants is completely used up, and then reaction stops. Such reactions are said to go to completion. The reaction between magnesium and oxygen is a good example of a reaction that goes to completion:

$$2Mg + O_2 \rightarrow 2MgO$$

Many other reactions, however, do not go to completion and are **reversible**. When the reactants and products have different colours, it is easy to demonstrate the reversibility of the reaction. For example, when dilute sulfuric

Notes

Other reactions which can be used in the laboratory to illustrate the reversibility of reactions include $[Co(H_2O)_6]^{2+}$(aq) with Cl^-(aq) and Fe^{3+}(aq) with SCN^-(aq).

acid is added to an aqueous solution containing yellow chromate(VI) ions, the following reaction occurs, forming orange dichromate(VI) ions:

$$2CrO_4^{2-} + 2H^+ \rightarrow Cr_2O_7^{2-} + H_2O$$

(yellow) (orange)

If an aqueous solution of sodium hydroxide is now added to the orange solution, the reaction is reversed and yellow chromate(VI) ions are re-formed:

$$Cr_2O_7^{2-} + 2OH^- \rightarrow 2CrO_4^{2-} + H_2O$$

(orange) (yellow)

The overall reaction can be represented by the equation:

$$2CrO_4^{2-} + 2H^+ \rightleftharpoons Cr_2O_7^{2-} + H_2O$$

(yellow) (orange)

The $\rightleftharpoons$ sign is used to indicate that the reaction is reversible. By convention, the reaction shown as occurring from left to right in the equation is called the **forward reaction** and the reaction occurring in the opposite direction is called the **backward reaction** or the **reverse reaction.** Since the reaction still continues in both directions, it is said to be **dynamic.** When both reactions occur at the same rate, the concentrations of the chromate(VI) and dichromate(VI) ions remain constant, and a **chemical equilibrium** has been established. A chemical equilibrium can only be established if reagents are neither added to, nor taken from, the reaction mixture.

When a chemical equilibrium is established:

- reactants and products are present at all times
- the reaction is dynamic, i.e. it proceeds in both directions
- the concentrations of reactants and products remain constant.

If reactants are added or if products are removed, the equilibrium is displaced.

Effects of changing reaction conditions

The effect on the equilibrium position of the following is considered below:

- change in concentration
- change in pressure
- change in temperature
- addition of a catalyst.

For most reactions the qualitative effect of changing reaction conditions can be predicted using **Le Chatelier's principle**.

Definition

Le Chatelier's principle *states that a system at equilibrium will respond to oppose any change imposed upon it.*

Effect of a change in concentration

If all other conditions remain the same and the concentration of any of the species involved in an equilibrium reaction is changed, then the concentrations of the other species must also change. Le Chatelier's principle can be used to deduce the changes that occur. For example, if the concentration of a reactant is increased, or the concentration of a product is decreased (e.g. by removing some of it), the position of equilibrium moves to the right and more product is formed.

Consider the reaction:

$$CH_3COOC_2H_5(l) + H_2O(l) \rightleftharpoons CH_3COOH(l) + C_2H_5OH(l)$$

If a little more water or $CH_3COOC_2H_5(l)$ is added, some of the additional reagent reacts and the equilibrium position is displaced to the right. As a result, the equilibrium yields of $CH_3COOH(l)$ and $C_2H_5OH(l)$ are both increased. Similarly, if more $CH_3COOH(l)$ or more $C_2H_5OH(l)$ is added, the equilibrium is displaced to the left and the equilibrium mixture will contain more $CH_3COOC_2H_5(l)$ and $H_2O(l)$.

Effect of a change in total pressure

Changes in total pressure have a significant effect on the composition of a mixture at equilibrium only if the reaction involves gases. The changes observed are due to changes in the concentrations of the species present. An increase in total pressure, according to Le Chatelier's principle, will displace the equilibrium in a direction that tries to reduce the increased pressure – the system responds by decreasing the number of moles of gas present, thus lowering the total pressure. The converse of this statement also applies.

For example, consider the reaction:

$$CH_4(g) + H_2O(g) \rightleftharpoons 3H_2(g) + CO(g)$$

In this equation, the total number of moles of gaseous reactants is two – i.e. 1 mol of $CH_4(g)$ plus 1 mol of $H_2O(g)$, and the total number of moles of gaseous products is four – i.e. 3 mol of $H_2(g)$ plus 1 mol of $CO(g)$.

Thus, at a given temperature, the equilibrium amount of products can be increased by *reducing* the total pressure, so that the system responds by moving to the right to produce a greater number of moles of gas, in an attempt to increase the pressure.

Essential Notes

If pressure is lowered, the rate of the reaction decreases.

Effect of a change in temperature

A change in temperature alters the rate of both the forward and the backward reactions. These are changed by different amounts, so the position of the equilibrium is altered. The simple rule which says that a system at equilibrium will react to oppose any change imposed upon it (Le Chatelier's principle) can be used to predict the effects of a change in temperature.

An increase in temperature is opposed by a movement of the position of equilibrium, either to the left or to the right, in order to absorb the added heat energy. Heat energy is absorbed by moving in the **endothermic** direction.

A decrease in temperature is opposed by a movement of the position of equilibrium, either to the left or to the right, in order to gain the lost heat energy. Heat energy is gained by moving in the **exothermic** direction.

Essential Notes

The converse is true if the temperature is *decreased.*

In an **exothermic reaction** heat energy is evolved. An increase in temperature requires removal of the added heat energy by an equilibrium shift in the endothermic direction. The equilibrium position is displaced to the left and the equilibrium mixture contains a lower concentration of products. It is important to note, however, that although the new equilibrium mixture obtained at a higher temperature contains less product, the time taken to reach this new equilibrium is reduced because of the increased rate of the reaction.

For example, consider the effect of a change in temperature on the exothermic reaction:

$$H_2(g) + I_2(g) \rightleftharpoons 2HI(g) \qquad \Delta H = -9.6\ \text{kJ mol}^{-1}$$

At 298 K, this equilibrium lies far to the right and the reaction mixture at equilibrium contains a high percentage of HI(g). If the temperature rises, the percentage of HI(g) present in the equilibrium mixture falls.

In an **endothermic reaction** heat energy is absorbed. An increase in temperature requires replacement of the lost heat energy, once again by an equilibrium shift in the endothermic direction. In this case, the equilibrium position is displaced to the right and the equilibrium mixture contains a higher concentration of products. Thus, for an endothermic reaction, an increase in temperature increases the equilibrium concentration of products.

Consider, for example, the endothermic reaction:

$$N_2(g) + O_2(g) \rightleftharpoons 2NO(g) \qquad \Delta H = 180\ \text{kJ mol}^{-1}$$

Essential Notes

Changes in pressure have no effect on the position of this equilibrium since the number of moles of gas on each side of the equation is the same.

At 298 K, the equilibrium lies so far to the left that the equilibrium mixture contains almost no NO(g). Increasing the temperature to 1500 K does increase the equilibrium yield of NO(g), but this is still too small for the direct combination of nitrogen and oxygen to be an economically viable method of preparing NO(g).

The effects of changes in temperature on equilibria can be summarised as follows: an *increase* in temperature always displaces the equilibrium in the *endothermic* direction (Table 11).

Table 11
Effects of temperature changes on equilibria

Reaction enthalpy	***Change in temperature***	***Displacement of equilibrium***	***Yield of product***	***Rate of attainment of equilibrium***
exothermic	increased	to the left	reduced	increased
exothermic	decreased	to the right	increased	reduced
endothermic	increased	to the right	increased	increased
endothermic	decreased	to the left	reduced	reduced

Effect of a catalyst

The addition of a catalyst to a mixture at equilibrium has no effect on the composition of the equilibrium mixture. This is because a catalyst causes an equal increase in the rates of both the forward and the backward reactions which themselves are equal at equilibrium. Hence, the equilibrium position is achieved more quickly, but the composition of the equilibrium mixture is unchanged.

Importance of equilibria in industrial processes

Many chemicals are manufactured on a large scale. The processes used are designed to give the optimum yield. All the factors which affect the position of a particular equilibrium reaction must be carefully considered.

Ethanol and methanol are important organic compounds which are increasingly being used as fuels. Both of these alcohols are manufactured on a large scale.

Ethanol production

Ethanol is produced industrially by the hydration of ethene:

$$C_2H_4(g) + H_2O(g) \rightleftharpoons C_2H_5OH(g) \qquad \Delta H = -46\ \text{kJ mol}^{-1}$$

Effect of pressure: The equation above shows that 2 mol of gaseous reactants (one of ethene and one of water) form 1 mol of gaseous product (ethanol). Application of Le Chatelier's principle predicts that a high pressure will favour the hydration of ethene. The pressure used is a compromise between the cost of generating high pressure and the additional value of the increased equilibrium yield. A typical pressure for this reaction is 6.5 MPa.

Effect of temperature: The hydration of ethene is an exothermic reaction. Application of Le Chatelier's principle predicts that this reaction will be opposed by an increase in temperature, so that the best equilibrium yield of ethanol will be obtained at low temperatures. However, at low temperatures the rate of reaction is slow and, although a high equilibrium yield might be achieved, it may take a long time to reach equilibrium. Increasing the temperature speeds up the rate of attainment of equilibrium, but reduces the equilibrium yield. A compromise between yield and speed of reaction is clearly necessary and a typical operating temperature for this reaction is around 300 °C.

Effect of a catalyst: The acid H_3PO_4 is used as a catalyst for this reaction. This catalyst increases the rate of both the forward and backward reactions to the same extent. Hence, the time taken to reach equilibrium is reduced, making the process more cost efficient, though the equilibrium yield of ethanol is unaltered.

Methanol production

Methanol is produced industrially by the reaction between carbon monoxide and hydrogen:

$$CO(g) + 2H_2(g) \rightleftharpoons CH_3OH(g) \qquad \Delta H = -90\ \text{kJ mol}^{-1}$$

Effect of pressure: This equation shows that 3 mol of gaseous reactants form 1 mol of gaseous product. Application of Le Chatelier's principle predicts that a higher equilibrium yield will be obtained at high pressure. Again, the cost of generating the high pressure must be balanced against the value of the increased equilibrium yield of methanol. A typical pressure used in industry is around 5 MPa.

Effect of temperature: Since the reaction between carbon monoxide and hydrogen is exothermic, application of Le Chatelier's principle predicts that a high equilibrium yield of methanol will be obtained at a low temperature. However, as the rate of the reaction will be low at a relatively low temperature, a compromise is again required and a typical operating temperature is around 400 °C.

Effect of a catalyst: The time required for the reaction to reach equilibrium is reduced by the use of a catalyst: a mixture of chromium(III) oxide, Cr_2O_3, and zinc oxide, ZnO, is often used as the catalyst for this reaction.

3.1.6.2 Equilibrium constant K_c for homogeneous systems

This section introduces a new concept – that of an equilibrium constant – and illustrates how this constant can be used:

- to provide a quantitative measure of the extent of reaction
- to determine the position of equilibrium.

Homogeneous system

A **homogeneous system** is one in which all the species present are in the *same phase*. In the case of equilibria, this usually means the liquid phase, although it also includes the possibility of homogeneous reactions in the gas phase.

The equilibrium constant K_c for a system at constant temperature

The **equilibrium constant K_c** is calculated from concentrations at constant temperature of the species involved in the equilibrium.

Definition

*The **equilibrium constant** for a reaction is obtained by multiplying together the concentrations of the products, each raised to the power of its coefficient in the stoichiometric equilibrium equation, and dividing this by the concentrations of the reactants, each also raised to the appropriate power.*

In general, for a reaction:

$$a\text{A(aq)} + b\text{B(aq)} \rightleftharpoons c\text{C(aq)} + d\text{D(aq)}$$

the equilibrium constant is:

$$K_c = \frac{[\text{C}]^c[\text{D}]^d}{[\text{A}]^a[\text{B}]^b}$$

where *a, b, c* and *d* are the numbers of moles of the species A, B, C and D which appear in the balanced equation for the equilibrium, and square brackets [ ] denotes a concentration in mol dm^{-3}.

Units of K_c

The units of K_c depend on the stoichiometry of the chosen equilibrium reaction. For example, the reaction:

$$2\text{A(aq)} + \text{B(aq)} \rightleftharpoons \text{C(aq)}$$

has $K_c = \dfrac{[\text{C}]}{[\text{A}]^2[\text{B}]}$ with units obtained by simplification:

$$\frac{\cancel{\text{mol dm}^{-3}}}{(\text{mol dm}^{-3})^2\,\cancel{(\text{mol dm}^{-3})}} = \frac{1}{(\text{mol dm}^{-3})^2} = \text{mol}^{-2}\,\text{dm}^{6}$$

If there are equal numbers of moles on both sides of the equilibrium equation, then K_c has no units.

Essential Notes

At equilibrium there is no net change in the concentrations of reactants and products, because any tendency for change in one direction is always balanced by an equal and opposite tendency for change in the other direction.

Essential Notes

homogeneous = same phase

Essential Notes

The *stoichiometric equation* is the balanced chemical equation for the reaction in question.

Essential Notes

a, b, c and *d* are called the **stoichiometric coefficients** of the balanced equilibrium equation.

Notes

K_c usually involves concentration terms expressed in the units mol dm^{-3}.

Note that the equilibrium constant K_c can also be deduced for a gas-phase reaction. The reaction:

$$2A(g) + B(g) \rightleftharpoons C(g) \text{ still has } K_c = \frac{[C]}{[A]^2[B]}$$

with [ ] representing the concentration of gaseous species in mol dm^{-3} of gas phase.

Because the expression for K_c involves the stoichiometric coefficients of the equilibrium equation, the numerical value of K_c, and also its units, are linked uniquely to the equation for which it is defined. Thus, the doubled equation above:

$$4A(g) + 2B(g) \rightleftharpoons 2C(g) \quad \text{has} \quad K_{c1} = \frac{[C]^2}{[A]^4[B]^2} = (K_c)^2$$

which has a value that is the square of the one for the previous equation.

By the same token, the equilibrium constant for the reverse reaction is the reciprocal of the original equilibrium constant:

$$C(g) \rightleftharpoons 2A(g) + B(g) \quad \text{has} \quad K_{c2} = \frac{[A]^2[B]}{[C]} = \frac{1}{K_c}$$

Calculations using K_c

The calculation of K_c is best illustrated by means of examples, as shown below.

Example

200 g of ethyl ethanoate and 7.0 g of water were refluxed together. At equilibrium, the mixture contained 0.25 mol of ethanoic acid.

Calculate the equilibrium constant for the hydrolysis of ethyl ethanoate.

Method

Determine the concentrations of all the species present using the equilibrium equation below.

Let the *initial* amount in moles of ethyl ethanoate be a, that of water b, and the total volume be V dm^3.

Let x mol each of ester and water react together forming x mol each of alcohol and acid, leaving $(a - x)$ mol of ester and $(b - x)$ mol of water.

Reaction:	$CH_3COOC_2H_5(l)$ +	$H_2O(l) \rightleftharpoons$	$CH_3COOH(l)$ +	$C_2H_5OH(l)$
Initial concn	a/V	b/V	0	0
Equilibrium concn	$(a - x)/V$	$(b - x)/V$	x/V	x/V

Calculation

Initial moles:

ethyl ethanoate: initial amount in moles $a = 200/88 = 2.27$ mol

water: initial amount in moles $b = 7/18 = 0.39$ mol

Equilibrium concentrations:

ethanoic acid = ethanol: amount in moles (given) x = 0.25 mol

ethyl ethanoate: amount in moles $(a - x)$ = (2.27 − 0.25) mol

water: amount in moles $(b - x)$ = (0.39 − 0.25) mol

Notes

The defining stoichiometric equation must always be stated before the equilibrium constant expression can be formulated.

Equilibrium concentrations:

$$[CH_3COOC_2H_5] = (2.27 - 0.25)/V = 2.02/V \text{ mol dm}^{-3}$$

$$[H_2O] = (0.39 - 0.25)/V = 0.14/V \text{ mol dm}^{-3}$$

$$[C_2H_5OH] = [CH_3COOH] = 0.25/V \text{ mol dm}^{-3}$$

Equilibrium constant:

$$K_c = \frac{[C_2H_5OH] \times [CH_3COOH]}{[CH_3COOC_2H_5] \times [H_2O]}$$

$$\text{Hence } K_c = \frac{(0.25/V)\text{ mol dm}^{-3} \times (0.25/V)\text{ mol dm}^{-3}}{(2.02/V)\text{ mol dm}^{-3} \times (0.14/V)\text{ mol dm}^{-3}} = 0.22 \text{ (no units)}$$

Comment

In this case K_c has no units; the concentration units cancel as there are equal amounts in moles on both sides of the equilibrium equation.

Example

2 mol of phosphorus(V) chloride vapour are heated to 500 K in a vessel of volume 20 dm^3. The equilibrium mixture contains 1.2 mol of chlorine. Calculate the value of the equilibrium constant K_c for the decomposition of phosphorus(V) chloride into phosphorus(III) chloride.

Method

Determine the concentrations of the three species present by using the equilibrium equation below.

Let x mol of PCl_5 decompose to form x mol each of PCl_3 and Cl_2.

Reaction:	$PCl_5(g)$	$\rightleftharpoons$	$PCl_3(g)$	+	$Cl_2(g)$
Initial concn	$2/V$		0		0
Equilibrium concn	$(2 - x)/V$		x/V		x/V

Calculation

Equilibrium moles:

$PCl_3 = Cl_2$: number of moles (given) x = 1.2 mol

PCl_5: number of moles $(2 - x)$ = $(2 - 1.2)$ mol

Equilibrium concentrations:

$[PCl_3] = [Cl_2]$: x/V = 1.2 mol/20 dm^3 = 0.06 mol dm^{-3}

$[PCl_5]$: $(2 - x)/V$ = $(2 - 1.2)$ mol/20 dm^3 = 0.04 mol dm^{-3}

Equilibrium constant:

$$K_c = \frac{[PCl_3] \times [Cl_2]}{[PCl_5]}$$

$$\text{Hence } K_c = \frac{0.06 \text{ mol dm}^{-3} \times 0.06 \text{ mol dm}^{-3}}{0.04 \text{ mol dm}^{-3}} = 0.09 \text{ mol dm}^{-3}$$

Comment

Because there are two moles of product but only one mole of reactant in the equation then, by cancellation, the units of K_c are mol dm^{-3}.

Qualitative effects of changes in temperature and concentration

The position of equilibrium and the value of the equilibrium constant

In section 3.1.6.1, the effects on the position of equilibrium of the following changes in conditions were considered, using Le Chatelier's principle.

- change in temperature
- change in concentration
- addition of a catalyst.

It is important to distinguish between the effects of these changes on postion of equilibrium with their possible effects on the value of the equilibrium constant, K_c.

Change in temperature

A change in temperature changes the value of the equilibrium constant K_c. According to Le Chatelier's principle, the constraint of higher temperature can be relieved if the equilibrium moves in the direction that *absorbs* some of the added heat, thus opposing the change in temperature.

Essential Notes

An increase in temperature will always increase the reaction rate and decrease the time required to reach equilibrium.

Exothermic reactions

In an **exothermic reaction** heat is given out as the reaction proceeds. This evolution of heat will tend to *raise* the temperature of the reaction mixture. An increase in temperature can be opposed by reaction in the direction which will absorb the added heat and so *decrease* the temperature. Thus, in an exothermic reaction the equilibrium is displaced to the *left* and the equilibrium mixture contains a *lower concentration of products.* The converse is true if the temperature is *decreased.*

Consider the effect of a change in temperature on the exothermic equilibrium reaction:

$$H_2(g) + I_2(g) \rightleftharpoons 2HI(g) \qquad \Delta H^\ominus = -9.6\ \text{kJ mol}^{-1}$$

for which the following values of K_c have been found:

Notes

For this reaction there are equal numbers of moles on both sides of the equilibrium equation, so the equilibrium constant has no units.

Temperature/K	Equilibrium constant, K_c
298	794
500	160
700	54

Table 12
Variation of the equilibrium constant with temperature for an exothermic reaction

K_c *decreases* with *increasing* temperature in an *exothermic* reaction.

Endothermic reactions

In an **endothermic reaction** heat is being taken in as the reaction proceeds. This absorbtion of heat will tend to *lower* the temperature of the reaction mixture. An increase in temperature can be opposed by reaction in the direction which will absorb the added heat so as to *decrease* the temperature. Thus, the equilibrium is displaced to the *right* and the equilibrium mixture contains a *higher concentration of products.* The converse is true if temperature is *decreased.*

Essential Notes

Increased temperature always shifts the equilibrium in the *endothermic direction.*

Decreased temperature always shifts the equilibrium in the *exothermic direction.*

The value of K_c is altered by changes in temperature.

Consider the effect of a change in temperature on the endothermic reaction:

$$N_2(g) + O_2(g) \rightleftharpoons 2NO(g) \qquad \Delta H^{\circ} = +180\ kJ\ mol^{-1}$$

for which the following values of K_c have been found:

Table 13
Variation of the equilibrium constant with temperature for an endothermic reaction

Temperature/K	Equilibrium constant, K_c
293	4×10^{-31}
700	5×10^{-13}
1500	1×10^{-5}

K_c *increases* with *increasing* temperature in an *endothermic* reaction.

The effects of changes in temperature on equilibria are summarised below:

Table 14
The effect of temperature on equilibrium

ΔH for reaction	Change in temperature	Shift of equilibrium	Yield of product	Equilibrium constant
exothermic	increase	to the left	reduced	reduced
exothermic	decrease	to the right	increased	increased
endothermic	increase	to the right	increased	increased
endothermic	decrease	to the left	reduced	reduced

Essential Notes

If the concentration of a reactant is increased (or the concentration of the product is decreased), the equilibrium opposes the change by moving to the right to give more product. The opposite happens if reactant is removed or product is added.

Change in concentration

At a given temperature the value of the equilibrium constant, K_c, is fixed.

If, at a given temperature, the concentration of any one species involved in the equilibrium is changed, then, as predicted by Le Chatelier's principle (see section 3.1.6.1) the concentration of all the other species will change so that the value of K_c remains constant.

If the reactants or products are gases, a change in the pressure of any gaseous species is equivalent to a change in the concentration of that species.

The example below illustrates the application of these principles to a real situation.

Example

Soda water is made by dissolving carbon dioxide in water. Suggest optimum conditions for the manufacture of soda water.

Method

The reaction involved is:

$$CO_2(g) \xrightleftharpoons{\text{water}} CO_2(aq) \qquad \Delta H^{\circ} \text{ is negative}$$

Answer

Soda water goes *flat* if warmed; the process is *exothermic* (heating causes it to go in the endothermic direction, forming more gas). So, cooling will be beneficial to making soda water. Increasing the pressure increases the concentration of the CO_2 gas, hence high pressures are helpful in making soda water.

Optimum conditions

Low temperature and high pressure, which agrees with common sense.

The effect of a catalyst on the equilibrium position and the equilibrium constant

Definition

*A **catalyst** alters the rate of a chemical reaction without itself being consumed.*

A catalyst does not have any effect on the position of equilibrium in a chemical reaction; hence it does not affect the value of the equilibrium constant. Thus a catalyst can *never* affect the yield of chemical processes. All that a catalyst can do is to *speed up the attainment of equilibrium.*

3.1.7 Oxidation, reduction and redox equations

The term **redox** is used for reactions which involve both **reduction** and **oxidation**. Originally the term oxidation was applied to the formation of a metal oxide when a metal reacted with oxygen. For example:

$$2Mg + O_2 \rightarrow 2MgO$$

The reverse of this reaction was given the name reduction, and reducing agents were substances which removed oxygen. Hence, in the reaction:

$$Fe_2O_3 + 2Al \rightarrow Al_2O_3 + 2Fe$$

aluminium behaves as a reducing agent. In the reaction:

$$CuO + C \rightarrow Cu + CO$$

carbon is the reducing agent.

The role of hydrogen as a reducing agent is also recognised in the following definition:

- oxidation – the addition of oxygen (or the removal of hydrogen)
- reduction – the removal of oxygen (or the addition of hydrogen).

Now consider the reactions:

$$SO_2 + H_2O + HgO \rightarrow H_2SO_4 + Hg$$

$$SO_2 + 2H_2O + Cl_2 \rightarrow H_2SO_4 + 2HCl$$

In the first case, HgO, by losing oxygen to form Hg, has clearly acted as the oxidising agent. In the second case, since oxidation of SO_2(aq) to SO_3(aq) has again occurred, the oxidising agent must be chlorine. To understand why chlorine is able to act as an oxidising agent we need to introduce the concept of oxidation state and redefine what we mean by the terms oxidation and reduction.

Oxidation states

For a simple ion, the oxidation state is the charge on the ion:

Na^+, K^+, Ag^+	have an oxidation state of	+1
Mg^{2+}, Ca^{2+}, Ba^{2+}	have an oxidation state of	+2
F^-, Cl^-, I^-	have an oxidation state of	−1
O^{2-}, S^{2-}	have an oxidation state of	−2

Notes

Oxidation state can also be called oxidation number.

The **oxidation state** of the central atom in a complex ion (an ion consisting of several atoms) is the charge it would have if it existed as a solitary simple ion

Table 15
Data for assignment of oxidation states

Species	Oxidation state
elements not combined with others	0
oxygen in compounds, except peroxides	−2
hydrogen in compounds, except in metal hydrides	+1
hydrogen in metal hydrides	−1
Group 1 metals in compounds	+1
Group 2 metals in compounds	+2

Essential Notes

For combined oxygen in peroxides, each oxygen has an oxidation state of −1.

without bonds to other species. Table 15 gives some data which can be used to establish the oxidation state of an atom in a complex ion.

Calculation of the oxidation state of a combined element in an oxo-ion

Using the data given in Table 15, the oxidation state of an atom in a complex ion can be calculated. The principles used are that the sum of the oxidation states in a neutral compound is zero and that the sum of the oxidation states in an ion is equal to the overall charge of the ion. For example, the oxidation state of phosphorus in PO_4^{3-} is determined as follows.

The overall charge is −3, therefore:

oxidation state of phosphorus + (4 × oxidation state of oxygen) = −3

Hence: oxidation state of phosphorus − 8 = −3

Thus: oxidation state of phosphorus = +5

Some more examples of oxo-ion complexes are given in Table 16.

Notes

The sum of the oxidation states of all the atoms in any complex is equal to the overall charge of that species.

Table 16
Calculation of the oxidation state of some central atoms in oxo-ions

Species	Number of oxygen atoms	Total oxidation number due to oxygen	Overall charge on the ion	Oxidation state of central atom	Name of species
SO_4^{2-}	4	−8	−2	+6	sulfate
NO_3^-	3	−6	−1	+5	nitrate
ClO_3^-	3	−6	−1	+5	chlorate(V)
ClO^-	1	−2	−1	+1	chlorate(I)

Notes

The ending 'ate' means that the ion has a negative charge and contains oxygen.

Redox equations

Redox reactions can be summarised as shown:

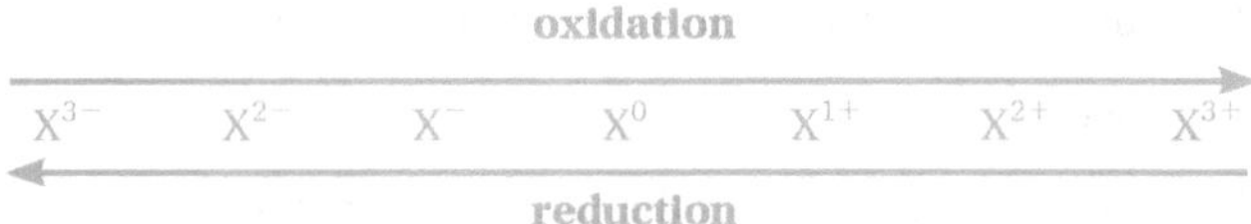

Notes

More precisely, SO_4^{2-} should be called sulfate(VI) and NO_3^- should be called nitrate(V) but these oxidation states are omitted in common usage.

Definitions

***Oxidation** is the process of electron loss. Oxidising agents are electron acceptors.*

***Reduction** is the process of electron gain. Reducing agents are electron donors.*

Identifying redox reactions

Redox reactions can readily be understood by the use of these definitions of oxidation and reduction.

In the reaction:

$$Fe_2O_3 + 2Al \rightarrow Al_2O_3 + 2Fe$$

the changes which occur are shown in the following half-equations:

$$Fe^{3+} + 3e^- \rightarrow Fe$$

$$Al \rightarrow Al^{3+} + 3e^-$$

The oxidation state of iron changes from +3 in Fe_2O_3 to zero in the uncombined metal, i.e. a reduction occurs. The oxidation state of aluminium changes from zero in the uncombined metal to +3 in Al_2O_3 and the aluminium metal is oxidised.

Equally, in the oxidation by chlorine of $SO_2(aq)$ to $SO_3(aq)$, i.e. $H_2SO_4(aq)$:

$$Cl_2 + SO_2 + 2H_2O \rightarrow H_2SO_4 + 2HCl$$

chlorine is reduced from oxidation state zero to oxidation state −1, as shown by the following half-equation:

$$Cl_2 + 2e^- \rightarrow 2Cl^-$$

The oxidation state of sulfur is increased from +4 to +6 by oxidation:

$$SO_2 + 2H_2O \rightarrow H_2SO_4 + 2H^+ + 2e^-$$

Example

Explain why the following is a redox reaction:

$$Mg + 2HCl \rightarrow MgCl_2 + H_2$$

Answer

In this reaction, the oxidation state of hydrogen, combined in hydrochloric acid, changes from +1 to zero in the uncombined element; i.e. the hydrogen ion is reduced by the magnesium metal. The uncombined magnesium metal, in oxidation state zero, is changed into combined magnesium in magnesium chloride, with an oxidation state of +2. Thus, the magnesium metal is oxidised by the hydrochloric acid.

The use of an ionic equation makes this redox reaction easy to recognise:

$$Mg + 2H^+ \rightarrow Mg^{2+} + H_2$$

Example

Explain why the following is a redox reaction:

$$MnO_2 + 4HCl \rightarrow MnCl_2 + Cl_2 + 2H_2O$$

Answer

Deductions using the oxidation states of combined oxygen and chlorine show that, in this reaction, manganese is reduced from oxidation state +4 to +2 by the chloride ions in hydrochloric acid. Some of the chlorine, combined in HCl, is converted into the uncombined element chlorine.

Notes

The mnemonic O I L R I G, which stands for Oxidation Is Loss, Reduction Is Gain, is a useful aid when learning how electrons are transferred in a redox reaction.

Essential Notes

Half-equations are discussed more fully on page 52.

This change, from oxidation state −1 to zero, is due to oxidation by manganese(IV) oxide. This simple ionic equation shows the changes in oxidation state:

$$Mn^{4+} + 2Cl^- \rightarrow Mn^{2+} + Cl_2$$

Example

Explain why the following is not a redox reaction:

$$MgO + 2HCl \rightarrow MgCl_2 + H_2O$$

Answer

At an initial glance this might appear to be a redox reaction as the magnesium oxide 'loses' oxygen. There is, however, no change in the oxidation state of any of the elements present. The reaction is that of an acid with a base, resulting in the formation of a salt and water.

$$MgO + 2H^+ \rightarrow Mg^{2+} + H_2O$$

Half-equations for redox reactions

Earlier in this section the equations:

$$Fe^{3+} + 3e^- \rightarrow Fe$$

$$Al \rightarrow Al^{3+} + 3e^-$$

were given. These are examples of **half-equations**. The overall equation for a redox reaction can be separated into two half-equations; one shows reduction, the other oxidation. In each case the half-equation is balanced using electrons, so that the overall charge on both sides of the half-equation is the same.

These equations are often much simpler than molecular equations because they only show the actual species involved in the reaction. It is only necessary to know the initial and final species in a redox reaction to be able to construct half-equations for both processes.

Construction of half-equations for reactions

When constructing any half-equation, the following points must be observed:

- only *one* element in a half-equation may change oxidation state
- the half-equation must balance for atoms
- the half-equation must balance for charge.

When constructing a half-equation for reactions occurring in aqueous solution, water provides a source of oxygen and any 'surplus' oxygen is converted into water by reaction with hydrogen ions from an acid.

By applying these rules, half-equations for the reduction of any species can be deduced.

Example

Deduce the half-equation for the reduction, in acid solution, of NO_3^- to NO.

Answer

The oxidation state of nitrogen changes from +5 in NO_3^- to +2 in NO and nitrogen is reduced. The oxidation state of the oxygen is still −2, but two of the three oxygens combine with four hydrogen ions (provided by the added acid) to form two water molecules:

$$NO_3^- + 4H^+ \rightarrow NO + 2H_2O$$

This incomplete half-equation now balances for atoms but not for charge, with a total charge of +3 on the left-hand side and zero on the right-hand side. Three electrons must be added to the left-hand side to give the balanced half-equation:

$$NO_3^- + 4H^+ + 3e^- \rightarrow NO + 2H_2O$$

Construction of overall equations for redox reactions

The overall equation for any redox reaction can be obtained by adding together two half-equations, making sure that the number of electrons given by the reducing agent exactly balances the number of electrons accepted by the oxidising agent.

Example

When chlorine gas is bubbled through an aqueous solution of potassium bromide, the solution turns yellow as bromide ions are oxidised to bromine by chlorine, which is itself reduced to chloride ions. Write half-equations for the oxidation of bromide ions and for the reduction of chlorine, and use these to deduce an overall equation for the reaction.

Answer

The half-equation for the reduction of chlorine is:

$$Cl_2 + 2e^- \rightarrow 2Cl^-$$

The half-equation for the oxidation of bromide ions to bromine is:

$$2Br^- \rightarrow Br_2 + 2e^-$$

The number of electrons gained by chlorine in the first equation must equal the number of electrons given by two bromide ions in the second equation. The overall equation can therefore be obtained by simply adding together the two half-equations. The overall equation does not involve electrons, because they cancel out:

$$\begin{array}{rcl} Cl_2 + 2e^- & \rightarrow & 2Cl^- \\ 2Br^- & \rightarrow & Br_2 + 2e^- \\ \hline Cl_2 + 2Br^- & \rightarrow & 2Cl^- + Br_2 \end{array}$$

Notes

Potassium ions take no part in this reaction and can be omitted from the equations. They are called spectator ions.

Essential Notes

Other useful examples include the reduction of concentrated H_2SO_4 to SO_2, S or H_2S when warmed with solid NaI (see this book, section 3.2.3.1).

Example

When concentrated nitric acid is added to copper metal, copper is oxidised to oxidation state +2 and nitric acid is reduced to nitrogen(IV) oxide. Write half-equations for the oxidation of copper and for the reduction of nitric acid, and use these to deduce an overall equation for the reaction.

Answer

The following half-equation for copper shows that uncombined copper metal is oxidised by loss of electrons:

$$Cu \rightarrow Cu^{2+} + 2e^-$$

The reduction reaction involves the reduction of nitrogen from +5 in HNO_3 to +4 in NO_2:

$$HNO_3 + H^+ + e^- \rightarrow NO_2 + H_2O$$

It is necessary to add H^+ ions to the left-hand side of this half-equation to combine with the 'surplus' oxygen (see page 52).

The two half-equations can now be combined to give an overall redox equation for the reaction. When this is done, the number of electrons required by the species being reduced and the number given by the species being oxidised must be the same, so that the equation for the overall redox reaction does not contain electrons.

In this example, the half-equation for the reduction of nitric acid must be doubled, so that the overall equation will correctly show that the two electrons lost by copper metal are accepted by nitric acid.

The overall equation for the reaction is given by addition:

$$Cu \rightarrow Cu^{2+} + 2e^-$$

$$2HNO_3 + 2H^+ + 2e^- \rightarrow 2NO_2 + 2H_2O$$

$$2HNO_3 + 2H^+ + Cu \rightarrow Cu^{2+} + 2NO_2 + 2H_2O$$

In certain cases, where the same species appears on both sides of an overall equation obtained by the addition of two half-equations, it is necessary to cancel such a species so that it only appears on one side of the final equation. Water molecules and hydrogen ions are the most common species which need to be treated in this way. Ions which take no part in the reaction are also omitted.

3.2 Inorganic chemistry

3.2.1 Periodicity

3.2.1.1 Classification

Elements are classified in the Periodic Table as shown in Fig 28.

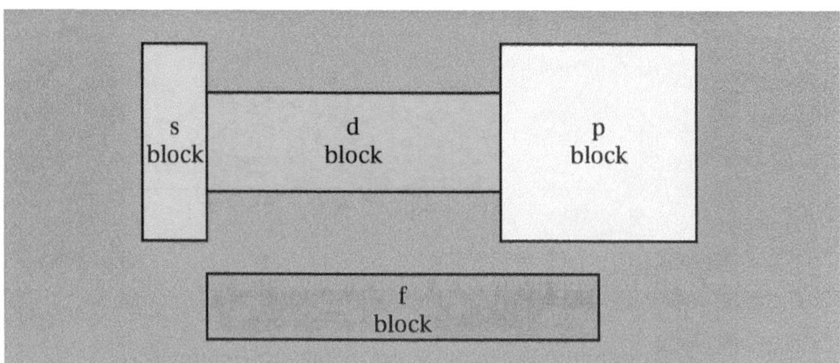

Fig 28
Classification of elements in the Periodic Table

In these blocks, the elements have their highest-energy electrons in s, p, d or f electronic sub-levels. For example:

- Li ($1s^2\,2s^1$) and Mg ($1s^2\,2s^2\,2p^6\,3s^2$) are **s-block elements**
- Cl ($1s^2\,2s^2\,2p^6\,3s^2\,3p^5$) is a **p-block element**
- Co ($1s^2\,2s^2\,2p^6\,3s^2\,3p^5\,4s^2\,3d^7$) is a **d-block element.**

Essential Notes

The letters s, p, d and f, which are used to describe electronic sub-levels (orbitals), originate from the description of lines in atomic emission spectra. The emission lines were described as **s**harp, **p**rincipal (the brightest lines), **d**iffuse and **f**aint (or fundamental).

3.2.1.2 Physical properties of Period 3 elements

Atomic radius

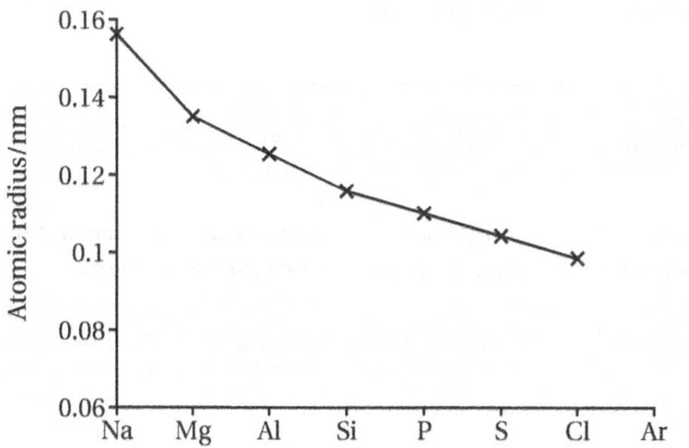

Fig 29
Atomic radii of the Period 3 elements

Essential Notes

The **atomic radius** of these elements is measured in suitable compounds of the element. Argon forms no compounds and therefore it is not possible to measure its atomic radius in a way which can be compared with the other elements.

The atomic radii decrease because, across the period, the nuclear charge increases and the outer electrons are attracted more strongly. They are therefore drawn closer to the nucleus, without any significant increase in shielding of the nuclear charge by the added electrons (see Fig 29).

First ionisation energy

This trend was discussed in this book, section 3.1.1.3.

Melting points

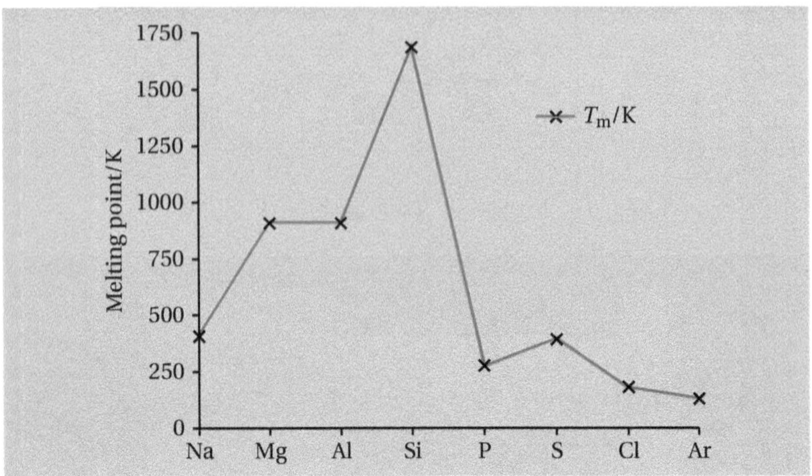

Fig 30
Melting points of the Period 3 elements

Essential Notes

Argon has a very low melting point because it exists as individual atoms. The atoms are not very polarisable and therefore the van der Waals' forces are very weak.

The variation in the melting points (Fig 30) is linked to the bond strength and the structure of the elements.

From sodium to aluminium the elements are metals. The melting points increase because, from sodium to aluminium, the atoms are smaller and have an increasing nuclear charge. Therefore, the strength of metallic bonding increases.

Silicon has a high melting point because it is macromolecular, with a diamond structure, and strong covalent bonds link all the atoms in three dimensions. A great deal of energy is required to break these bonds.

Phosphorus, sulfur and chlorine are all molecular substances. The melting point (Table 17) of each is determined by the strength of van der Waals' forces between molecules which, in turn, are determined by the sizes of the molecules. Each of these elements has a low melting point because the van der Waals' forces are weak and easily broken.

Table 17
Melting points of some Period 3 elements

Molecular formula	P_4	S_8	Cl_2
Melting point/K	317	392	172

The sulfur molecule is the biggest and most polarisable, so that sulfur has a melting point which is higher than that of phosphorus or chlorine.

Essential Notes

Phosphorus and sulfur exist in different forms, but the molecular formulas shown are the simplest and most common.

3.2.2 Group 2, the alkaline earth metals

The Group 2 elements are the metals in the second group of the Periodic Table. They are therefore s-block elements in which the outermost electrons are in a full s sub-shell.

Trends in physical properties

Some important physical properties of the Group 2 elements Mg–Ba are given in Table 18.

Element	*Electron configuration*	*Atomic radius/nm*	*First ionisation energy/kJ mol^{-1}*	*Melting point/K*
Mg	[Ne]3s^2	0.136	737	924
Ca	[Ar]4s^2	0.174	590	1116
Sr	[Kr]5s^2	0.191	549	1042
Ba	[Xe]6s^2	0.198	503	998

Table 18
Physical properties of the Group 2 elements Mg–Ba

Essential Notes

Remember, electron configurations are sometimes abbreviated by giving only the electrons beyond the previous noble gas.

Atomic radius

On descending the group from magnesium to barium, the atomic radii increase. This trend is due to the increasing number of electron shells, resulting in the outermost electrons being progressively further away from the nucleus (see Fig 31).

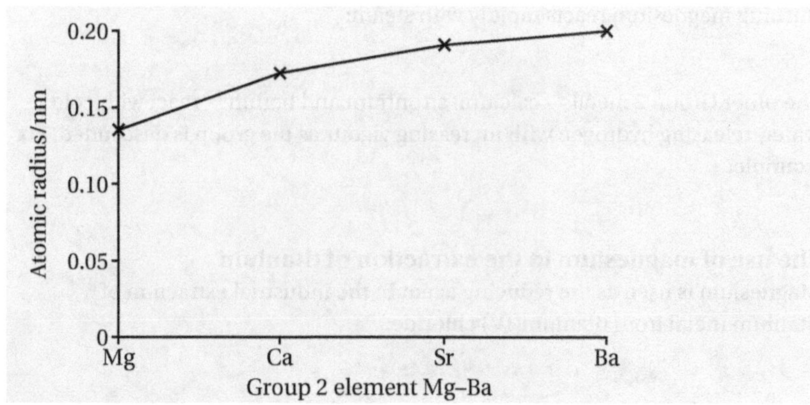

Fig 31
Atomic radius of the Group 2 elements Mg–Ba

First ionisation energy

The first ionisation energies of the elements decrease down the group as the atomic radius increases and the outermost electrons become increasingly shielded from the positive charge of the nucleus.

Melting points

The Group 2 elements are metals with high melting points (see Fig 32).
In metallic structures, the positive ions (cations) (M^{2+} in this group) are surrounded by a 'sea' of outermost electrons. There are two of these delocalised

electrons for each M^{2+} ion. As the metal ions become larger going down the group, the strength of the metallic bonds generally decreases, because the decreasing density of charge (or charge-to-size ratio) of the ions means that there is less attraction for the delocalised electrons. The expected general decrease in melting point from calcium to strontium can be seen. Magnesium has a lower melting point than expected because it has a different crystal structure from the other metals.

Essential Notes

Magnesium has a hexagonal close-packed structure, calcium and strontium have face-centred cubic structures and barium has a body-centred cubic structure.

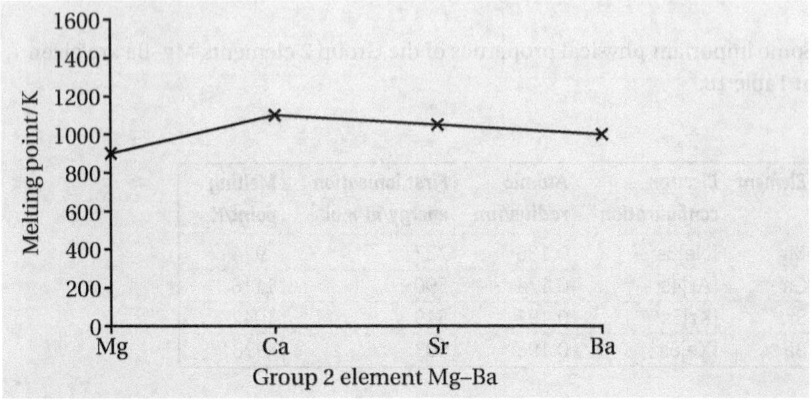

Fig 32 Melting points of the Group 2 elements Mg–Ba

Trends in chemical properties

Reactions with water

The reactivity of the Group 2 elements with water increases on descending the group. Magnesium reacts only very slowly with cold water:

$$Mg(s) + H_2O(l) \rightarrow Mg(OH)_2(s) + H_2(g)$$

Burning magnesium reacts rapidly with steam:

$$Mg(s) + H_2O(g) \rightarrow MgO(s) + H_2(g)$$

The other Group 2 metals - calcium, strontium and barium - react with cold water, releasing hydrogen with increasing vigour as the group is descended. For example:

$$Ca(s) + 2H_2O(l) \rightarrow Ca(OH)_2(aq) + H_2(g)$$

Notes

The carrying out of simple test-tube reactions to identify Group 2 ions is a required practical activity.

The use of magnesium in the extraction of titanium

Magnesium is used as the reducing agent in the industrial extraction of titanium metal from titanium(IV) chloride:

$$TiCl_4(l) + 2Mg(l) \rightarrow Ti(s) + 2MgCl_2(s)$$

The magnesium chloride by-product is removed from the titanium by vacuum distillation at high temperature.

The solubility of hydroxides

The solubility of the hydroxides of Group 2 elements increases on descending the group. Magnesium hydroxide is only sparingly soluble in water. Calcium hydroxide dissolves to form 'lime water', but is not as soluble as strontium hydroxide, while barium hydroxide is soluble in water and produces strongly alkaline solutions:

$$Ba(OH)_2(s) + aq \rightarrow Ba^{2+}(aq) + 2OH^-(aq)$$

Uses of magnesium and calcium hydroxides

Magnesium hydroxide is used as an antacid for the relief of indigestion caused by an excess of acid in the stomach:

$$Mg(OH)_2(s) + 2H^+(aq) \rightarrow Mg^{2+}(aq) + 2H_2O(l)$$

Magnesium hydroxide also acts as a laxative and is used to relieve constipation. It is taken orally either as chewable tablets or in a suspension often called 'milk of magnesia'.

Calcium hydroxide is used by farmers to reduce soil acidity, so that a wider range of crops can be grown, and to provide calcium ions which are essential for plant growth. The 'lime' used by farmers is often a mixture of calcium carbonate together with calcium hydroxide.

The use of calcium oxide and calcium carbonate to remove sulfur dioxide from flue gases

Sulfur dioxide is a toxic gas and, being soluble in water, can cause **acid rain** by forming a solution of sulfurous acid, H_2SO_3. High in the atmosphere, ultra-violet radiation provides the energy for sulfur dioxide to react with oxygen to form sulfur trioxide. Sulfur trioxide is very soluble in water and forms sulfuric acid, which also occurs in acid rain.

A lot of sulfur dioxide is produced by the burning of fuels in power stations. However, this pollutant is not released into the atmosphere, but is removed from the gases passed up the chimney (flue) by a process called flue-gas desulfurisation.

Several alkaline substances can be used to remove the acidic sulfur dioxide. Some methods use calcium oxide (quicklime) which is easily obtained by heating calcium carbonate (limestone). The product of the reaction of calcium oxide and sulfur dioxide is calcium sulfite, $CaSO_3$:

$$CaO + SO_2 \rightarrow CaSO_3$$

Calcium carbonate is also used:

$$CaCO_3 + SO_2(g) \rightarrow CaSO_3 + CO_2$$

The solubility of sulfates

The solubility of the sulfates of Group 2 elements decreases from the soluble magnesium sulfate to the insoluble barium sulfate.

In the laboratory, the insolubility of barium sulfate is used as a test for the presence of sulfate ions in solution. Dilute acid (hydrochloric or nitric) and a solution of barium ions (from barium chloride or barium nitrate) are added to the solution under test and the appearance of a white precipitate indicates the presence of sulfate ions:

$$Ba^{2+}(aq) + SO_4^{2-}(aq) \rightarrow BaSO_4(s)$$

In the absence of acids, carbonate ions interfere with this test because barium carbonate is also a white solid, insoluble in water. The carbonate ions are removed by adding either hydrochloric or nitric acid. The acid reacts with the carbonate ions to generate carbon dioxide gas, thereby removing them from solution and preventing the precipitation of barium carbonate:

$$2H^+(aq) + CO_3^{2-}(aq) \rightarrow CO_2(g) + H_2O(l)$$

Use of barium sulfate in medicine

Because barium sulfate blocks X-rays and is not toxic (as it is so insoluble in water or body fluids), it is used in medicine to aid the investigation of problems in digestive and bowel systems. An aqueous suspension of barium sulfate is taken orally, if an image of the digestive system is required, or injected through the anus, if images of the bowel region are needed.

3.2.3 Group 7(17), the halogens

The halogens form a family of non-metallic elements which show clear similarities and well-defined trends in their properties as the relative atomic mass increases.

3.2.3.1 Trends in properties

It is not necessary to remember the data, but it is necessary to know and be able to explain the observed trends. Some data are given in Table 19.

Table 19
Group 7 data

Element	Atomic number	Outer electrons	Atomic radius/nm	Radius of X^- ion/nm	Boiling point/K	Electro-negativity
F	9	$2s^2 2p^5$	0.071	0.133	85	4.0
Cl	17	$3s^2 3p^5$	0.099	0.180	238	3.0
Br	35	$4s^2 4p^5$	0.114	0.195	332	2.8
I	53	$5s^2 5p^5$	0.133	0.215	457	2.5

Trend in electronegativity of the halogens

The term electronegativity was introduced in this book, section 3.1.3.6. **Electronegativity** is the power of an atom to attract electron density in a covalent bond. The data show that the electronegativity of the halogens decreases as the atomic number increases.

To explain this trend, three important factors must be considered:

- the **atomic number**, which gives the nuclear charge. As this increases, the attraction for the bonding electron pair in a covalent bond might be expected to increase.
- the number of **electron shells**, which is indicated by the outer electron configuration of the atom. As this number increases, the shielding of the outer electrons from attraction by the nucleus increases. This results in the outer electrons being less strongly attracted.
- the **atomic radius** of the atom. The attraction between oppositely charged particles falls rapidly as the distance between them increases. As the radius of the atom increases, the outer electrons are further from the nucleus, which therefore attracts them less strongly.

The electronegativity of an element depends on a balance between these three factors. The changes in the values of electronegativity given for the halogens show that the increase in shielding and the increase in atomic radius more than compensate for the increase in nuclear charge.

The large electronegativity value for fluorine means that the bond between an element and fluorine is likely to be more polar than the bond formed between the same element and the other halogens.

Trend in boiling point of the halogens

All halogens exist as diatomic molecules, X_2. The attraction between these molecules in the liquid state is due to weak intermolecular forces called van der Waals' forces (see this book, section 3.1.3.7). These are caused by temporary fluctuations in electron density within the molecules, resulting in temporary dipole attractions between the molecules.

The magnitude of van der Waals' attractive forces increases with the size of the molecules. This fact explains why, as both atomic and molecular radii of the halogens increase with increasing atomic number, the boiling points of the halogens also increase, as shown by the data in Table 19.

Trends in chemical properties

Oxidising power of the halogens

When any reagent is oxidised, electrons are taken from it. The electrons are accepted by the oxidising agent, which is itself reduced. The trend in oxidising power of the halogens is characterised by a decrease from strongly-oxidising fluorine to weakly-oxidising iodine.

The reasons for the decreasing trend in oxidising power from fluorine to iodine down Group 7 are quite complex, in that they involve an overall balance of energies in the following process:

$$\tfrac{1}{2}X_2 + e^- \rightarrow X^-$$

It is helpful to consider this process as occurring in three stages:

- the strength of the X—X bond (breaking to form X atoms in the gaseous phase)
- the affinity of an X atom for an electron (forming an X^- ion in the gaseous phase)
- the energy released when the X^- ion goes into solution or into a crystal lattice.

These three features affect the halogens differently:

- the very strong oxidising ability of the fluorine molecule can be attributed partly to the weakness of the F—F bond
- the electron affinity does not vary greatly from one halogen atom to the next, so has little effect on the relative oxidising power
- the fluoride ion, being the smallest, has the most to gain from being hydrated or entering a crystal, whereas the iodide ion, being the largest, benefits much less.

Thus, the trend is a decrease in oxidising power from fluorine to iodine:

$$F_2 > Cl_2 > Br_2 > I_2$$

The relative oxidising power of chlorine, bromine and iodine can be determined experimentally in the laboratory by a series of displacement reactions. In these experiments, aqueous solutions of the three halogens are added separately to aqueous solutions containing the other two halide ions. The results are given in Tables 20–22.

Notes

Fluorine is the strongest of all oxidising agents.

Notes

The energy released when the X^- ion goes into solution or becomes part of an ionic lattice decreases as the size of the ion increases, being largest for the small F^- ion and least for the big I^- ion.

Notes

The F—F bond is weak because of repulsion between non-bonding electron pairs on the two atoms. The electron pairs are close together because the F atoms are small.

Essential Notes

Sea water contains a low concentration of bromide ions. Bromine is extracted from it by treating the sea water with chlorine. The liberated bromine is expelled from the water using air, and is then concentrated in a series of separate stages.

Table 20
The reactions of Cl_2(aq) with Br^-(aq) and I^-(aq)

Halide	*Observations*	*Conclusion*	*Equation*
Br^-(aq)	yellow/brown solution	Br_2 displaced	$2Br^- + Cl_2 \rightarrow 2Cl^- + Br_2$
I^-(aq)	brown solution and/or black precipitate	I_2 displaced	$2I^- + Cl_2 \rightarrow 2Cl^- + I_2$

Essential Notes

Iodine is almost insoluble in water, but in the presence of iodide ions it dissolves to form the complex ion I_3^-(aq) which is brown.

Table 21
The reactions of Br_2(aq) with Cl^-(aq) and I^-(aq)

Halide	*Observations*	*Conclusion*	*Equation*
Cl^-(aq)	no change	Cl_2 not displaced	no reaction
I^-(aq)	brown solution and/or black precipitate	I_2 displaced	$2I^- + Br_2 \rightarrow 2Br^- + I_2$

Table 22
The reactions of I_2(aq) with Cl^-(aq) and Br^-(aq)

Halide	*Observations*	*Conclusion*	*Equation*
Cl^-(aq)	no change	Cl_2 not displaced	no reaction
Br^-(aq)	no change	Br_2 not displaced	no reaction

Essential Notes

Fluorine is far too dangerous to be used other than in specially equipped laboratories by specially-trained staff. However, experiments using fluorine have been carried out and show that fluorine will oxidise all other halide ions to the halogen.

These results confirm the order of oxidising power by showing that:

- chlorine will displace bromine and iodine
- bromine will displace iodine but not chlorine
- iodine will not displace either chlorine or bromine.

Trends in properties of the halides

Trends in the reducing ability of the halide ions

The ability of halogen molecules to behave as oxidising agents by accepting additional electrons to form halide ions was considered above. When a halide ion behaves as a reducing agent, it loses an electron to the reagent it is reducing; this process is the reverse of that in which a halogen molecule acts as an oxidising agent. The trend in reducing power of the halide ions shows a decrease from the strongly-reducing iodide ion to the non-reducing fluoride ion:

$$I^- > Br^- > Cl^- > F^-$$

Reactions of sodium halides with sulfuric acid

The trend in the reducing power of the halide ions is shown in the reaction of solid halide salts with concentrated sulfuric acid. The oxidation state of sulfur in sulfuric acid is +6. This can be reduced to +4, 0 or −2 depending on the reducing power of the halide ion. Experimental results are given in Table 23.

NaX	Observations	Products	Type of reaction
NaF	steamy fumes	HF	acid–base (F^- acting as a base)
NaCl	steamy fumes	HCl	acid–base (Cl^- acting as a base)
NaBr	steamy fumes	HBr	acid–base (Br^- acting as a base)
	colourless gas	SO_2	redox (reduction product of H_2SO_4)
	brown fumes	Br_2	redox (oxidation product of Br^-)
NaI	steamy fumes	HI	acid–base (I^- acting as a base)
	colourless gas	SO_2	redox (reduction product of H_2SO_4)
	yellow solid	S	redox (reduction product of H_2SO_4)
	smell of bad eggs	H_2S	redox (reduction product of H_2SO_4)
	black solid, purple fumes	I_2	redox (oxidation product of I^-)

Table 23
The reactions of concentrated sulfuric acid with solid sodium halides

Essential Notes

These reactions can be demonstrated in a laboratory fume cupboard but full safety precautions must be taken. Hydrogen fluoride is an extremely dangerous gas and, in the presence of water, will even etch glass.

These results indicate that:

- iodide ions can reduce the sulfur in H_2SO_4 from oxidation state +6 to +4, as SO_2, then to 0, as the element sulfur, and finally to −2, as H_2S
- bromide ions can reduce the sulfur in H_2SO_4 from oxidation state +6 to +4, as SO_2
- fluoride and chloride cannot reduce the sulfur in H_2SO_4 under these conditions.

Notes
Deriving equations for the reactions which occur provides valuable revision of redox reactions.

The use of silver nitrate solution to identify and distinguish between halide ions

Silver fluoride is soluble in water but silver chloride, silver bromide and silver iodide are all insoluble. Silver chloride, bromide and iodide are precipitated when an aqueous solution containing the appropriate halide ion is treated with an aqueous solution of silver nitrate. Dilute nitric acid is added to the solution under test before addition of silver nitrate solution to prevent the formation of other insoluble compounds, such as Ag_2CO_3. The colours of the three silver salts formed with chloride, bromide and iodide ions, and their different solubilities in aqueous ammonia, can be used as a test for the presence of the halide. These results are summarised in Table 24.

Notes
If you are asked to describe an observation, you must always link a colour to a solution or a solid (precipitate).

Halide	Precipitate	Observation	Solubility of precipitate in ammonia solution
F^-	none	–	–
Cl^-	AgCl	white solid	soluble in dilute $NH_3(aq)$
Br^-	AgBr	cream solid	sparingly soluble in dilute $NH_3(aq)$, soluble in concentrated $NH_3(aq)$
I^-	AgI	yellow solid	insoluble in concentrated $NH_3(aq)$

Table 24
Testing for halide ions using $AgNO_3(aq)$ and $NH_3(aq)$

These results show that the solubility of the silver halides in ammonia solution decreases in the following order:

AgCl > AgBr > AgI

Notes
The carrying out of simple test-tube reactions to identify halide ions is a required practical activity.

The addition of ammonia solution to the silver halide precipitate formed is used to eliminate any potential confusion caused by the similar colours of these precipitates. The different solubilities of AgCl, AgBr and AgI in ammonia lead to a clear identification of the halide ion originally present.

3.2.3.2 Uses of chlorine and chlorate(I)

The products obtained when chlorine reacts with water depend on the conditions used. Under normal laboratory conditions, a very pale green solution is formed, showing the presence of the element chlorine, and an equilibrium is established:

$$Cl_2 + H_2O \rightleftharpoons HCl + HClO$$

This reaction is an example of a **disproportionation** reaction in which one species, in this case chlorine, is simultaneously both oxidised and reduced:

$$\begin{array}{lcccc} \text{Oxidation state of chlorine:} & 0 & & -1 & +1 \\ & Cl_2 + H_2O & \rightleftharpoons & HCl + & HClO \end{array}$$

If universal indicator is added to a solution of chlorine water, it first turns red since both the reaction products are acids, i.e. hydrochloric acid, HCl, which is a strong (fully ionised) acid, and chloric(I) acid, HClO, which is a weak (slightly ionised) acid. The red colour then disappears and a colourless solution is left because chloric(I) acid is a very effective bleach.

Essential Notes

Because this reaction is rather slow, it is best to leave an inverted test tube containing chlorine water in sunlight for several days, after which sufficient oxygen will have been produced to give a positive test with a glowing splint.

If chlorine is bubbled through water in the presence of bright sunlight, or the green solution of chlorine water is left in bright sunlight, a colourless gas is produced and the green colour, due to chlorine, fades. Tests show that the colourless gas evolved is oxygen. Under these conditions, chlorine oxidises water to oxygen and is itself reduced to chloride ions:

$$2Cl_2 + 2H_2O \rightarrow 4H^+ + 4Cl^- + O_2$$

Water treatment

Chlorine and chlorine compounds are used in water treatment. For many years, small quantities of chlorine have been added to drinking water and to swimming pools in order to kill disease-causing bacteria. In drinking water, the major public health hazards are due to bacteria that cause cholera and typhus. In swimming pools, the dangerous bacteria killed by chlorine are often types of *E. coli* that originate from human waste.

The decision about the amount of chlorine to be added to drinking water supplies is a good illustration of how society has to assess the advantages and disadvantages of using chemicals to sterilise water supplies. Too little chlorine is ineffective in killing bacteria, but too much damages the health of consumers. Clearly, the health benefits of carefully controlled chlorination of drinking water outweigh the potential toxic effects. More recently, fluoridation of the water supply to improve dental health has required similar assessment of the advantages and disadvantages.

The concentration of chlorine in drinking water is approximately 0.7 mg dm^{-3}. Higher concentrations are used in swimming pools. Great care is taken to ensure that the correct amounts of chlorine are used because chlorine itself is very toxic. In addition, chlorine can react with organic waste material in water to form organochlorine compounds which may be toxic. Although it is well known that failure to chlorinate water results in serious health risks, there is little information to support the claim that the formation of organochlorine compounds in water is a long-term health risk.

Reaction of chlorine with cold dilute aqueous sodium hydroxide

When chlorine reacts with cold water, an equilibrium is established between the reactants and the two acidic products:

$$Cl_2 + H_2O \rightleftharpoons HCl + HClO$$

If water is replaced by cold dilute sodium hydroxide, the effect is to displace the equilibrium to the right as the hydroxide ions react with the acids produced:

$$Cl_2 + 2NaOH \rightarrow NaCl + NaClO + H_2O$$

or $Cl_2 + 2OH^- \rightarrow Cl^- + ClO^- + H_2O$

This reaction is of great commercial importance because the mixture of sodium chloride and sodium chlorate(I) is used as a bleach.

Practical and mathematical skills

In the AS paper 1, approximately 15% of marks will be allocated to the assessment of skills related to practical chemistry. A minimum of 20% of the marks will be allocated to assessing level 2 mathematical skills. These practical and mathematical skills are likely to overlap.

The required practical activities assessed in this paper are:

- Make up a volumetric solution and carry out a simple acid–base titration
- Measurement of an enthalpy change
- Carry out simple test-tube reactions to identify cations (Group 2 and NH_4^+) and anions (halide, hydroxide, carbonate and sulfate ions).

The practical skills assessed in the paper are:

1. Independent thinking

Examination questions may require problem solving and the application of scientific knowledge and understanding in practical contexts. For example, a question may ask how, in a novel context, an experiment could be carried out to determine the concentration of a solution or to evaluate an enthalpy change.

2. Use and application of scientific methods and practices

This skill may be assessed by asking for critical comments on a given experimental method. Questions may ask for conclusions from given observations: for example, in the reactions of Group 2 cations and halide anions. Questions may require the presentation of data in appropriate ways, such as in tables or graphs. It will also be necessary to express numerical results – for example, from titration and enthalpy determinations – to an appropriate precision with reference to uncertainties and errors in volumetric apparatus and in thermometer readings.

3. Numeracy and the application of mathematical concepts in a practical context

There is some overlap between this skill and the use and the application of scientific methods and practices. Questions may require the plotting and interpretation of graphs. For example, in experiments to determine an accurate enthalpy change, it may be necessary to understand how to extrapolate a cooling curve in order to determine an accurate value of temperature at a certain point in an experiment. Level 2 mathematical skills will be required to analyse practical data. Examples of such skills will be met in mole calculations that involve either the volume of a solution or the volume of a gas in cm^3 and in the determination of equilibrium constants.

4. Instruments and equipment

It will be necessary to know and understand how to use volumetric equipment, including graduated flasks, burettes and pipettes. Other instruments and equipment met in this paper are those required to determine enthalpy changes and to make appropriate observations of test-tube reactions. Questions will assess the ability to understand in detail how to ensure that the use of this laboratory equipment leads to results that are as accurate as possible.

The mathematical skills assessed in this paper are:

1. **Arithmetic and numerical computation**

- **Recognise and make use of appropriate units in calculations.**

  All numerical answers should be given with the appropriate units. Questions may require conversions between units: for example, cm^3 to dm^3 and J to kJ.

- **Recognise and use expressions in decimal and standard form.**

  When required, it will be necessary to express answers to an appropriate number of decimal places and to carry out calculations and express answers in ordinary or standard form. For example, calculations involving concentrations, equilibrium constants and Avogadro's constant (6.022×10^{23}) may involve numbers in standard form and conversions between standard and ordinary form.

- **Use ratios fractions and percentages.**

  Examples of this skill include percentage yields, atom economies and balancing equations.

- **Estimate results.**

  Calculations of this type could include the evaluation of how a change in concentration of one component in an equilibrium mixture might affect the yield of product.

- **Use calculators.**

  The ability to use calculators to handle numbers in standard form (for example, Avogadro's constant) may be assessed.

2. **Handling data**

- **Use an appropriate number of significant figures.**

  Understand that a calculated result can only be reported to the limits of the least accurate measurement: for example, the temperature change in an enthalpy experiment.

- **Find arithmetic means.**

  These may be required, for example, in the determination of relative atomic mass from isotopic masses and abundances and also in the selection of appropriate readings to determine a mean titre.

- **Identify uncertainties in measurements and when data are combined.**

  It will be necessary to demonstrate an ability to determine uncertainty when two burette readings are used to calculate a titre value.

3. **Algebra**

- **Change the subject of an equation.**

  For example, in order to calculate the temperature of a gas from the ideal gas equation $pV = nRT$, the equation should be rearranged into $T = pV / nR$

- **Substitute numerical values into algebraic equations using appropriate units.**

  For example, in the equation $E_k = \frac{1}{2}mv^2$ in a time of flight mass spectrometer, using E_k in J, m in kg and v in m s^{-1}

- **Solve algebraic equations.**

  For example, to calculate the speed of an ion in m s^{-1} from the equation $E_k = \frac{1}{2}mv^2$.

4. **Graphs**

- **Plot two variables from experimental data.**

  For example, in an experiment to determine an enthalpy change, plot temperature versus time graphs from experimental data and then extrapolate the plots to determine the temperatures before and after mixing the reactants.

5. **Geometry and trigonometry**

- **Use angles and shapes in regular 2D and 3D structures.**

  Questions may assess an ability to predict, identify and sketch the shapes of bond angles in simple molecules and ions with and without lone pairs: for example NH_3, AlF_6^{3-}.

Practice exam-style questions

1 **(a)** Complete the following table.

Name of fundamental particle	Relative mass	Relative charge
	5.45×10^{-4}	
		+1
		0

6 marks

(b) Give the meaning of the term *mass number.*

1 mark

(c) In terms of the numbers of fundamental particles, explain the difference between two isotopes of the same element.

2 marks

(d) Give the full chemical symbol for the isotope that has seven electrons and six neutrons in one atom.

2 marks

(e) Give the number of protons, neutrons and electrons in the ion $^{31}P^{3-}$

3 marks

(f) Give the full electronic configuration, including sub-shells, of the following species.

(i) Mg^{2+}

(ii) Cr

(iii) S^{2-}

(iv) Fe^{2+}

4 marks

(g) Use the Avogadro constant, 6.02214×10^{23} mol^{-1}, and your knowledge about relative masses of isotopes to calculate the mass of one atom of carbon-12. Give your answer in kilograms in standard form to the appropriate number of significant figures.

2 marks

Total marks: 20

2 A schematic diagram of a time of flight mass spectrometer is shown below.

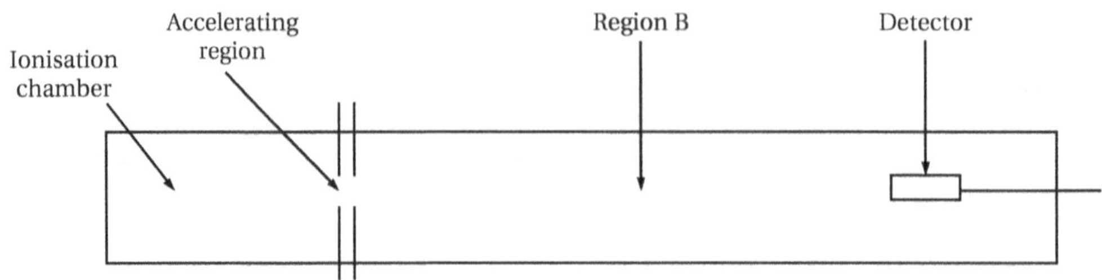

(a) State how acceleration is achieved and state the common characteristic of the ions after acceleration.

How achieved ______________________________

Common characteristic ______________________________ 2 marks

(b) State what occurs in the part of the mass spectrometer labelled Region B.

______________________________ 1 mark

(c) Explain how the ions are detected and how the relative abundance of each ion is determined.

____

____

______________________________ 3 marks

(d) A sample of an element from a meteorite is analysed in a mass spectrometer. The relative masses and relative abundances of the isotopes are shown in the table below.

Isotope mass	54	56	57
Relative abundance	10.8	91.7	0.544

(i) Calculate the relative atomic mass of the element in the sample. Give your answer to 1 d.p.

____

____

______________________________ 3 marks

(ii) Identify the element.

______________________________ 1 mark

(iii) Explain why the relative atomic mass of the element in the sample is different from that in the Periodic Table.

______________________________ 1 mark

(iv) State which of the isotopes would have the shortest time of flight in the mass spectrometer. Explain your answer.

____

____

______________________________ 2 marks

Total marks: 13

3 A 2.00 g sample of an alloy of magnesium (alloyed with only one other metal, **M**) was reacted completely with 50.0 cm^3 (an excess) of a hydrochloric acid solution; 949 cm^3 of hydrogen gas were formed at 293 K and 101 kPa. The excess of hydrochloric acid was neutralised by reaction with exactly 21.3 cm^3 of 1.00 mol dm^{-3} sodium hydroxide solution. You may assume that metal M does not react with hydrochloric acid.

$$Mg(s) + 2HCl(aq) \rightarrow MgCl_2(aq) + H_2(g)$$

$$HCl(aq) + NaOH(aq) \rightarrow NaCl(aq) + H_2O(aq)$$

(a) State the ideal gas equation and use it to calculate the number of moles of hydrogen gas produced.

3 marks

(b) Use your answer to part (a) to calculate the mass of magnesium metal in the sample of magnesium alloy.

2 marks

(c) Calculate the number of moles of NaOH used to neutralise the excess of HCl.

2 marks

(d) Use your answers to parts (a) and (c) to calculate the moles of HCl present in the original 50 cm^3 of solution, and hence calculate the concentration of HCl in this solution.

4 marks

(e) Use your answer to part (b) to calculate the percentage by mass of metal **M** in the 2.00 g sample of alloy.

2 marks

(f) **(i)** Explain why it is necessary to assume that the metal **M** does not react with hydrochloric acid.

1 mark

(ii) You are provided with a pure sample of metal **M**. Describe a simple chemical test to show that metal **M** does not react with hydrochloric acid.

2 marks

Total marks: 16

4 Sodium chloride and graphite have different electrical conductivities. State and explain their conductivities in terms of bonding and structure.

6 marks

5 **(a)** What is meant by the term *enthalpy change?*

2 marks

(b) Define the term *standard enthalpy of formation.*

3 marks

(c) **(i)** Use the mean bond enthalpy data given in the table below to calculate the enthalpy change for the following reaction.

$$C_3H_8(g) + 5O_2(g) \rightarrow 3CO_2(g) + 4H_2O(g)$$

	C—C	C—H	O=O	C=O	O—H
Mean bond enthalpy/kJ mol^{-1}	348	412	496	743	463

3 marks

(ii) The enthalpy change for this reaction can also be calculated by using the standard enthalpies of formation of the reactants and the products.

Explain why the value obtained by using mean bond enthalpies is likely to be less accurate than the value obtained from enthalpies of formation.

2 marks

(iii) State how the enthalpy change for the reaction given in (c) (i) would differ if the water formed in the reaction were a liquid rather than a gas. Explain your answer.

How enthalpy change would differ

Explanation

2 marks

Total Marks: 12

6 (a) State and explain the trend in the first ionisation energies of the Group 2 elements from Mg to Ba.

Trend ______________________________

Explanation ______________________________

______________________________ 3 marks

(b) Magnesium has a higher melting point than sodium. Explain this observation in terms of bonding and structure.

____

____

______________________________ 3 marks

(c) Magnesium and barium both react with water. State the conditions under which they react and write separate equations for the reactions.

____

____

____

____

____

______________________________ 4 marks

(d) Calcium oxide and calcium hydroxide are used in industry.

(i) Write an equation for the conversion of calcium oxide into calcium hydroxide.

Equation ______________________________ 1 mark

(ii) Explain why calcium hydroxide is important in agriculture.

______________________________ 1 mark

(iii) Calcium oxide is used to remove sulfur dioxide from flue gases in power plants that burn fossil fuels. Give an equation for the reaction.

______________________________ 1 mark

(iv) Suggest the type of reaction in parts (d) (ii) and (d) (iii).

______________________________ 1 mark

(e) Barium sulfate is used in medicine even though reference books state that barium compounds are very toxic.

Give a medical use of barium sulfate and explain why this barium compound is not toxic.

Use of barium sulphate ______________________________

Explanation ______________________________

______________________________ 2 marks

Total marks: 16

7 **(a)** State and explain the trend in boiling points in Group 7 from fluorine to iodine.

3 marks

(b) When chlorine is added to water a green solution is formed. When sodium hydroxide is added to this solution the green colour fades.

(i) Give the formula of the species responsible for the green colour.

1 mark

(ii) Write an equation for the reaction of chlorine with water in the dark and use your equation and *Le Chatelier's principle* to explain why the addition of sodium hydroxide causes the green colour to fade.

3 marks

(c) Solid sodium halides react with concentrated sulfuric acid in different ways.

(i) Give the formula of a sodium halide that does **not** undergo a redox reaction with concentrated sulfuric acid. Write an equation for the reaction which takes place.

Formula

Equation 2 marks

(ii) Give the formula of a sodium halide that will reduce concentrated sulfuric acid to hydrogen sulfide. Write a half-equation for the reduction of sulfuric acid to hydrogen sulphide.

Formula

Equation 2 marks

(d) Give a reagent which you could use to distinguish AgCl(s) from AgBr(s) and state the observations you would make.

Reagent

Observation with AgCl(s)

Observation with AgBr(s) 3 marks

Total marks: 14

8 Give a reagent or a combination of reagents and any necessary conditions which could be used to distinguish between the following pairs of compounds. Give any observations and write an ionic equation for any reactions which occur.

(a) $MgCl_2(aq)$ and $BaCl_2(aq)$

Reagent ______________________________

Observation with $MgCl_2(aq)$ ______________________________

Observation with $BaCl_2(aq)$ ______________________________

Equation ______________________________ 4 marks

(b) $Na_2SO_4(aq)$ and $(NH_4)_2SO_4(aq)$

Reagent and conditions ______________________________

Observation with $Na_2SO_4(aq)$ ______________________________

Observation with $(NH_4)_2SO_4(aq)$ ______________________________

Equation ______________________________ 4 marks

(c) $Na_2SO_4(aq)$ and $NaNO_3(aq)$

Reagent ______________________________

Observation with $Na_2SO_4(aq)$ ______________________________

Observation with $NaNO_3(aq)$ ______________________________

Equation ______________________________ 4 marks

(d) $CaCl_2(aq)$ and $CaI_2(aq)$

Reagent ______________________________

Observation with $CaCl_2(aq)$ ______________________________

Observation with $CaI_2(aq)$ ______________________________

Equation ______________________________ 4 marks

Total marks: 16

9 This question is about the determination of the enthalpy of combustion of propan-1-ol, $CH_3CH_2CH_2OH$, using the apparatus shown in the diagram.

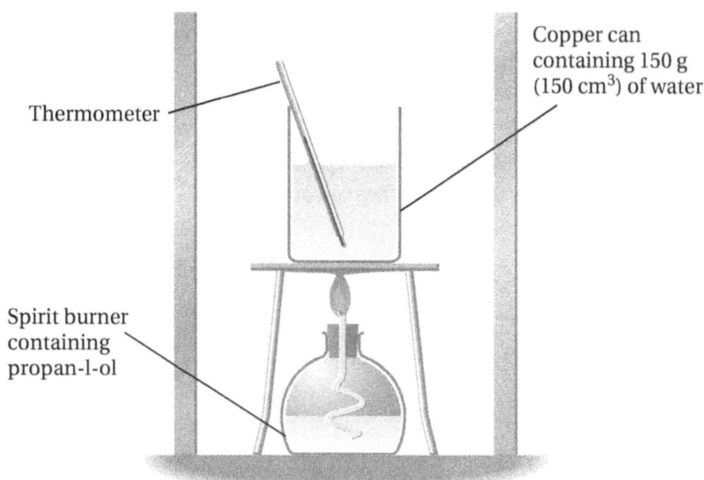

In one experiment, the water temperature increased from 18.2 °C to 35.4 °C and the mass of the burner decreased from 15.46 g to 14.58 g.

The specific heat capacity of water is 4.18 J g^{-1} K^{-1}.

(a) Calculate the enthalpy of combustion of propan-1-ol in this experiment. Give your answer to the appropriate number of significant figures

5 marks

(b) One major source of error in this determination is incomplete combustion. Give the other major source of error and suggest two improvements to the method in order to reduce this error.

3 marks

(c) A student suggested that the percentage uncertainty was greatest in measuring the mass change. Use the data to decide whether the student was correct. Suggest how this percentage uncertainty could be reduced and justify your answer.

The mass of water was measured with a measuring cylinder with an uncertainty of $\pm 1.0\ cm^3$.

The temperature change was measured with a thermometer with a total uncertainty of ±0.2 °C.

The mass change was measured with a balance with a total uncertainty of ±0.02 g.

5 marks

Total marks: 13

10 This question is about oxidation and reduction.

(a) In terms of electrons, state what is meant by

(i) Oxidation

(ii) Reducing agent 2 marks

(b) State the oxidation state of the nitrogen in each of the following species

(i) N_2

(ii) NH_3

(iii) $H_2NO_3^+$ 3 marks

(c) Aluminium can convert nitrate(V) ions, NO_3^-, in acid conditions, into ammonium ions, NH_4^+, and is itself converted into aluminium ions, Al^{3+}.

(i) Construct a half-equation for the conversion of aluminium atoms into aluminium ions.

1 mark

(ii) Construct a half-equation for the conversion of nitrate(V) ions into ammonium ions in acid conditions.

1 mark

(iii) Combine the two half-equations to give an overall equation for the reaction of nitrate(V) ions in acidic conditions with aluminium.

1 mark

Total marks: 8

Multiple choice questions

1 The time of flight of ions, t, can be shown to be related to the mass, m, and charge, q, by the equation:

$$t = k\sqrt{\left(\frac{m}{q}\right)}$$

where k is a constant which depends only on the spectrometer and the operating conditions.

In a given spectrometer, an ion of $^{48}Ti^{+}$ has a time of flight of 32 microseconds.

Under the same conditions, an ion of $^{24}Mg^{2+}$ would have a time of flight of:

A 16 microseconds

B 32 microseconds

C 45 microseconds

D 64 microseconds

2 Choose the letter which corresponds to the incorrect statement:

A Magnesium has a higher third ionisation energy than aluminium.

B CH_4 has a lower boiling point than SiH_4.

C Boron trifluoride has a permanent dipole.

D Hydrazine, H_2NNH_2, has hydrogen bonding between its molecules.

3 Ammonia is formed according to the following equation:

$$N_2(g) + 3H_2(g) \rightleftharpoons 2NH_3(g)$$

At a given temperature, K_c has a value of 0.11 $dm^6\ mol^{-2}$. In an experiment, the concentration at equilibrium of hydrogen is 0.16 mol dm^{-3} and the concentration of nitrogen is 0.12 mol dm^{-3}.

The concentration of ammonia, in mol dm^{-3}, is:

A 5.4×10^{-5}

B 7.4×10^{-3}

C 1.4×10^{2}

D 1.8×10^{4}

4 Choose the letter which corresponds to the incorrect statement:

A Arsine, AsH_3, has a bond angle of approximately 107°.

B The BrF_4^- ion has four bonding pairs and two lone pairs.

C Concentrated aqueous ammonia can distinguish solid samples of silver chloride and silver bromide.

D A catalyst increases the rate of attainment of an equilibrium but it does not affect the value of the equilibrium constant, K_c, for the reaction.

Answers

Question	Answer		Marks
1 (a)	electron (1) relative charge –1 proton (1) relative mass 1 (1) neutron (1) 1 (1)	(1)	6
1 (b)	number of protons + number of neutrons in the nucleus of an atom	(1)	1
1 (c)	same number of protons and electrons different number of neutrons	(1) (1)	2
1 (d)	${}^{13}_{7}N$ 13, 7 (1) N (1)		2
1 (e)	15 protons 16 neutrons 18 electrons	(1) (1) (1)	3
1 (f) (i) (ii) (iii) (iv)	$1s^22s^22p^6$ $1s^22s^22p^63s^23p^64s^23d^4$ (or $3d^5\ 4s^1$) $1s^22s^22p^63s^23p^6$ $1s^22s^22p^63s^23p^63d^6$	(1) (1) (1) (1)	4
1 (g)	Mass of one mole of carbon 12 is exactly 12.0000 g = 1.20000×10^{-2} kg Mass of one atom = mass of one mole/Avogadro number = $1.20000 \times 10^{-2}/6.02214 \times 10^{23} = 1.99264713 \times 10^{-26} = 1.99265 \times 10^{-26}$ kg (to 6 s.f.)	(1) (1)	2
			Total 20
2 (a)	*How achieved* electric field *Common characteristic* have same kinetic energy	(1) (1)	2
2 (b)	ion drift	(1)	1
2 (c)	The ions hit the detector and gain an electron this causes a current in the detector the current is proportional to the abundance of the ions	(1) (1) (1)	3
2 (d) (i)	relative atomic mass = $\frac{(54 \times 10.8 + 56 \times 91.7 + 57 \times 0.544)}{(10.8 + 91.7 + 0.544)}$ = 55.7 (3 s.f.)	(1) (1) (1)	3
2 (d) (ii)	iron	(1)	1
2 (d) (iii)	There are different isotopes/different relative abundances of the isotopes in the sample from those on earth.	(1)	1
2 (d) (iv)	the one with relative mass of 54 smaller the mass the shorter the time of flight	(1) (1)	2
			Total 13
3 (a)	$PV = nRT$ $n = PV/RT = 101\,000 \times 949 \times 10^{-6}/(8.31 \times 293)$ = 0.0394	(1) (1) (1)	3
3 (b)	moles Mg = 0.0394 mass = moles × 24.3 = 0.957	(1) (1)	2

Question	Answer		Marks
3 (c)	moles = 21.3 × 1.00/1000 = 0.0213	(1) (1)	 2
3 (d)	moles of HCl reacted = 2 × 0.0394 = 0.0788 moles in excess = 0.0213 total = 0.1001 concentration = 0.1001 × (1000/50.0) = 2.00 mol dm^{-3}	(1) (1) (1) (1)	 4
3 (e)	mass of M = 2.00 − 0.957 = 1.043 g % = (1.04/2.00) × 100 = 52.2%	(1) (1)	 2
3 (f) (i)	impossible to determine moles of Mg if M also reacts	(1)	1
3 (f) (ii)	weigh then place in HCl and allow to react then wash, dry and re-weigh	(1) (1)	 2
			Total 16
4	**This answer is marked using levels of response.**		
	Level 3: 5–6 marks All parts are covered and the explanation of each part is generally correct and virtually complete. Answer communicates the whole process coherently and shows a logical progression from part 1 and part 2 to overall explanation.		
	Level 2: 3–4 marks All parts are covered but the explanation of each part may be incomplete OR two parts are covered and the explanations are virtually complete. Answer is mainly coherent and shows a progression. Some statements may be out of order and incomplete.		
	Level 1: 1–2 Two parts are covered but the explanation of each part may be incomplete and contain inaccuracies OR only one part is covered but the explanation is mainly correct and is virtually complete. Answer includes some isolated statements but there is no attempt to present them in a logical order or show confused reasoning.		
	Level 0: 0		
	Bonding and structure sodium chloride has ionic bonding with giant ionic lattice graphite covalent bonding and macromolecular structure		
	Conductivity sodium chloride insulator when solid good conductor when liquid graphite good conductor as solid		
	Explanation sodium chloride ions free to move in liquid but not in solid graphite has delocalised electrons		

Question	Answer		Marks
5 (a)	heat energy evolved at constant pressure	(1) (1)	 2
5 (b)	the enthalpy change when one mole of a substance is formed from its elements with all reactants and products in their standard states under standard conditions	(1) (1) (1)	 3
5 (c) (i)	ΔH_r = Σ enthalpy of bonds broken – Σ enthalpy of bonds formed = [(2 × 348) + (8 × 412) + (5 × 496)] – [(6 × 743) + (8 × 463)] = –1690 kJ mol^{-1}	(1) (1) (1)	 3
5 (c) (ii)	mean bond enthalpies are values averaged over many compounds so mean bond enthalpies are not exact for specific compounds	(1) (1)	 2
5 (c) (iii)	the value would show the reaction to be more exothermic heat is evolved when molecules attract each other, changing from a gas to a liquid	(1) (1)	 2
			Total 12
6 (a)	*Trend* decreases *Explanation* more shells more shielding (so attraction for outer electron to the nucleus decreases down the group)	(1) (1) (1)	 3
6 (b)	both metallic lattices magnesium ions have a greater charge than sodium ions greater attraction between the ion and the delocalised electrons in magnesium	(1) (1) (1)	 3
6 (c)	Magnesium needs to be heated in steam $Mg + H_2O \rightarrow MgO + H_2$ Barium reacts with cold water $Ba + 2H_2O \rightarrow Ba(OH)_2 + H_2$	(1) (1) (1) (1)	 4
6 (d) (i)	$CaO + H_2O \rightarrow Ca(OH)_2$	(1)	1
6 (d) (ii)	neutralises acid soil	(1)	1
6 (d) (iii)	$CaO + SO_2 \rightarrow CaSO_3$	(1)	1
6 (d) (iv)	neutralisation	(1)	1
6 (e)	*Use of barium sulfate* as a barium meal/in X-rays of the digestive system *Explanation* barium sulfate is insoluble so barium ions are not absorbed	(1) (1)	 2
			Total 16
7 (a)	*trend*: increases *explanation*: atoms increase in size/have more electrons greater van der Waals' forces between molecules	(1) (1) (1)	 3
7 (b) (i)	Cl_2	(1)	1
7 (b) (ii)	$Cl_2 + H_2O \rightleftharpoons HCl + HOCl$ sodium hydroxide reacts with the acid and lowers the concentration equilibrium moves to the right to restore concentration of acid concentration of chlorine decreases and so colour fades	(1) (1) (1)	 3
7 (c) (i)	*Formula* NaF or NaCl *Equation* $NaCl + H_2SO_4 \rightarrow NaHSO_4 + HCl$	(1) (1)	 2

Question	Answer		Marks
7 (c) (ii)	*Formula* NaI *Equation* $H_2SO_4 + 8H^+ + 8e^- \rightarrow H_2S + 4H_2O$	(1) (1)	2
7 (d)	*Reagent* dilute aqueous ammonia *Observation with AgCl(s)* dissolves *Observation with AgBr(s)* no visible change	(1) (1) (1)	3
			Total 14
8 (a)	see table below		4
8 (b)	*Reagent and conditions* NaOH and heat *Observation with $Na_2SO_4(aq)$* no visible change *Observation with $(NH_4)_2SO_4(aq)$* pungent fumes/alkaline gas *Equation* $2NH_4^+(aq) + 2OH^-(aq) \rightarrow 2NH_3(g) + H_2O(l)$	(1) (1) (1) (1)	4
8 (c)	*Reagent* $BaCl_2/Ba(NO_3)_2$ *Observation with $Na_2SO_4(aq)$* white ppte *Observation with $NaNO_3(aq)$* no visible change *Equation* $Ba^{2+}(aq) + SO_4^{2-}(aq) \rightarrow BaSO_4(s)$	(1) (1) (1) (1)	4
8 (d)	*Reagent* silver nitrate *Observation with $CaCl_2(aq)$* white ppte *Observation with $CaI_2(aq)$* yellow ppte *Equation* $Ag^+(aq) + Cl^-(aq) \rightarrow AgCl(s)$ or $Ag^+(aq) + I^-(aq) \rightarrow AgI(s)$	(1) (1) (1) (1)	4
			Total 16
9 (a)	energy change $= mc\,\Delta T$ $= 150 \times 4.18 \times 17.2$ J $= 10\,784.4$ J Moles of propan-1-ol = mass/M_r = 0.88/60.0 = 0.0147 mol $\Delta H_{combustion}$ = heat change/moles = $-10\,784.4/0.0147$ J mol^{-1} $= -740$ kJ mol^{-1} (2 s.f.)	(1) (1) (1) (1) (1)	5
9 (b)	heat loss use a lid/use draught shields/lag the sides of the calorimeter (any two)	(1) (2)	3
9 (c)	percentage uncertainty in measuring water = 1.0/150 × 100 = 0.67% percentage uncertainty in measuring temperature = 0.2/17.2 × 100 = 1.2% percentage uncertainty in measuring mass = 0.02/0.88 × 100 = 2.3% so the student was correct use a more precise balance/measure mass to 3 d.p. or use more water/have higher temperature rise so the mass change of spirit burner is greater	(1) (1) (1) (1) (1)	5
			Total 13

8 (a) table:

Reagent (1)	observation with $MgCl_2(aq)$ (1)	observation with $BaCl_2(aq)$ (1)	ionic equation (1)
H_2SO_4/Na_2SO_4	no visible change	white ppte	$Ba^{2+}(aq) + SO_4^{2-}(aq) \rightarrow BaSO_4(s)$
NaOH	white ppte	no visible change	$Mg^{2+}(aq) + 2OH^-(aq) \rightarrow Mg(OH)_2(s)$

Question	Answer		Marks
10 (a) (i)	loss of electrons	(1)	1
10 (a) (ii)	donates electrons to another species	(1)	1
10 (b) (i)	N_2 0	(1)	1
10 (b) (ii)	NH_3 −3	(1)	1
10 (b) (iii)	$H_2NO_3^+$ +5	(1)	1
10 (c) (i)	$Al \rightarrow Al^{3+} + 3e^-$	(1)	1
10 (c) (ii)	$NO_3^- + 10H^+ + 8e^- \rightarrow NH_4^+ + 3H_2O$	(1)	1
10 (c) (iii)	$8Al + 3NO_3^- + 30H^+ \rightarrow 8Al^{3+} + 3NH_4^+ + 9H_2O$	(1)	1
			Total 8
Multiple choice questions			
1 A			
2 C			
3 B			
4 C			

The table below highlights aspects of *mathematical and practical skills* in the exemplar questions.

Question	Mathematical skill
1 g	MS 0.0, MS 0.1, MS 0.4, MS 1.1
2 d	MS 1.2
3 *passim*	MS 0.0, MS 0.2, MS 2.2, MS 2.3
5 c	MS 2.4
9 a i	MS 0.0, MS 1.1
9 c	MS 1.3
MCQ 1	MS 0.2
MCQ 3	MS 0.4, MS 2.2, MS 2.3
MCQ 4	MS 4.1

Question	Practical skill
5 c	PS 1.1
9 a	PS 1.1
9 b	PS 2.1
9 c	PS 2.3

Glossary

acid rain	contains quantities of carbonic, nitric and sulfuric acids
allotropes	different structural modifications of an element
atomic (proton) number (*Z*)	the number of protons in the nucleus of an atom
atomic radius	half the distance between the nuclei of atoms of the same element when linked by a single covalent bond or in a metallic crystal
Avogadro constant (*L*)	6.022×10^{23} mol^{-1}
backward (or reverse) reaction	one that goes from right to left in an equation
bond dissociation enthalpy	the enthalpy change for the breaking of a covalent bond, with all species in the gaseous state
calorimeter	apparatus used to measure heat change
catalyst	a substance that alters the rate of a reaction without itself being consumed
chemical equilibrium	the point at which, in a reversible reaction, both the forward and backward reactions occur at the same rate, with the concentrations of all reactants and products remaining constant
concentration	$\frac{\text{number of moles of solute}}{\text{volume of solution in dm}^3}$ with units mol dm^{-3}
co-ordinate bond	a covalent bond formed when the pair of electrons originates from one atom
covalent bond	a shared pair of electrons
dative covalent bond	a covalent bond formed when the pair of electrons originates from one atom
d-block element	an element that has its highest-energy electron in a d sub-shell
delocalised electrons	electrons that are spread over many ions in a metal lattice or many atoms in graphite and which are free to move
dispersion forces	the weakest forces of attraction that exist between atoms or molecules. These forces result when electrons on adjacent atoms are displaced and induce temporary dipoles (also known as van der Waals' forces or London forces)
disproportionation	a reaction in which the same species is simultaneously oxidised and reduced
dynamic equilibrium	one which proceeds simultaneously in both directions
electronegativity	the power of an atom to attract the shared electrons in a covalent bond
electron shells	energy levels into which electrons are distributed
empirical formula	the simplest ratio of atoms of each element in a compound
endothermic	the gain of heat energy by a system; the enthalpy change is positive
endothermic reaction	one in which heat energy is taken in
energy levels	the specific values of energy that an electron may have in an atom

enthalpy change (ΔH)	the amount of heat energy released or absorbed when a chemical or physical change occurs at constant pressure
enthalpy of fusion	the enthalpy required to change one mole of a solid into a liquid, i.e. $X(s) \rightarrow X(l)$
enthalpy of vaporisation	the enthalpy required to change one mole of a liquid into a gas, i.e. $X(l) \rightarrow X(g)$
equilibrium constant (K_c)	the ratio of concentrations of products and reactants raised to the powers of their stoichiometric coefficients; e.g. for the reaction $3A \rightleftharpoons 2B + C$ $\quad K_c = \frac{[B]^2[C]}{[A]^3}$
exothermic	the loss of heat energy by a system; the enthalpy change is negative
exothermic reaction	one in which heat energy is given out
f-block element	an element that has its highest-energy electron in an f sub-shell
first ionisation energy	the enthalpy change for the removal of one mole of electrons from one mole of atoms of an element in the gas phase, i.e. $X(g) \rightarrow X^+(g) + e^-$
first law of thermodynamics	energy can be neither created nor destroyed, but can only be converted from one form into another
forward reaction	one that goes from left to right in an equation
full equation	a balanced symbol equation with the formulas of all reagents on the left-hand side and the formulas of all the products on the right-hand side
giant ionic lattice	see *ionic crystal*
giant metallic lattice	see *metallic crystal*
half-equation	a balanced equation for an oxidation or a reduction that shows a species losing or gaining electrons, e.g. for the process $SO_4^{2-} + 4H^+ + 2e^- \rightarrow SO_2 + 2H_2O$
Hess's law	the enthalpy change of a reaction depends only on the initial and final states of the reaction and is independent of the route by which the reaction occurs
homogeneous system	one with all species present in the same phase
hydrogen bonding	an intermolecular force between the lone pair on an electronegative atom (N, O or F) and a hydrogen atom bonded to such an electronegative atom
ideal gas	one that obeys the ideal gas equation, $pV = nRT$
ion	an atom or group of atoms which has lost or gained one or more electrons, giving it a positive or negative charge
ionic bond	the electrostatic force of attraction between oppositely charged ions
ionic crystal	a lattice of positive and negative ions bound together by electrostatic attractions
ionic equation	a simplified version of a balanced symbol equation, showing only the ions which are actively involved in the reaction
isotopes	atoms with the same number of protons but different numbers of neutrons
Le Chatelier's principle	a system at equilibrium will respond to oppose any change imposed on it

London forces	the weakest forces of attraction that exist between atoms or molecules. These forces result when electrons on adjacent atoms are displaced and induce temporary dipoles (also known as van der Waals' forces or dispersion forces)
macromolecular (giant covalent) crystal	a giant, covalently bonded lattice structure
macromolecule	a giant molecule with a regular three- or two-dimensional lattice structure
mass number (A)	the total number of protons and neutrons in the nucleus of one atom of the element
mean bond enthalpy	the average of several values of the bond dissociation enthalpy for a given type of bond, taken from a range of different compounds
metallic bond	electrostatic attraction between metal ions and delocalised electrons
metallic crystal	a lattice of metal ions surrounded by delocalised electrons
mole	the name mole is given to the amount of substance: 1 mol of particles/entities is 6.022×10^{23} particles/entities
molecular crystal	a lattice of covalent molecules held together by weak intermolecular forces
molecular formula	the formula that gives the actual number of atoms of each element in a molecule
neutral atoms	contain an equal number of positively charged protons and negatively charged electrons
octahedral	the spatial arrangement with one atom at the centre of six other atoms, with four atoms in its plane, one atom above this plane and one atom below this plane
orbitals	volumes in space around the nucleus within which electrons are most likely to be found
oxidation	the process of electron loss
oxidation state (number)	the charge a central atom in a complex ion would have if it existed as a solitary simple ion without bonds to other species
p-block element	an element that has its highest-energy electron in a p sub-shell
percentage atom economy	$\frac{\text{mass of desired product}}{\text{total mass of reactants}} \times 100$ it is a measure of how much of a desired product in a reaction is formed from the reactants
permanent dipole–dipole force	attraction between the slightly positive end of one polar molecule and the slightly negative end of an adjacent polar molecule
percentage yield	$\frac{\text{actual mass of product}}{\text{maximum theoretical mass of product}} \times 100$ it is a practical measure of the efficiency of a reaction
polarity	the displacement of electron density (formation of an electric dipole) in a covalent bond, or in a molecule, due to a difference in electronegativity
redox	used for reactions that involve both reduction and oxidation

reduction	the process of electron gain
relative atomic mass (A_r)	$\frac{\text{average mass of one atom of an element}}{\frac{1}{12} \times \text{the mass of one atom of } ^{12}\text{C}}$
relative molecular mass (M_r)	$\frac{\text{average mass of one molecule}}{\frac{1}{12} \times \text{the mass of one atom of } ^{12}\text{C}}$
reversible reaction	one which does not go to completion but can occur in either direction
s-block element	an element that has its highest-energy electron in an s sub-shell
second ionisation energy	the enthalpy change for the removal of one mole of electrons from one mole of unipositive ions in the gas phase, i.e. $X^+(g) \rightarrow X^{2+}(g) + e^-$
standard conditions	usually taken as 100 kPa and 298 K
standard enthalpy of combustion ($\Delta_c H^\ominus$)	the enthalpy change, under standard conditions, when 1 mol of a substance is burned completely in oxygen, with all reactants and products in their standard states
standard enthalpy of formation ($\Delta_f H^\ominus$)	the enthalpy change, under standard conditions, when 1 mol of a compound is formed from its elements, with all reactants and products in their standard states
standard state	the normal, stable state of an element or compound under standard conditions, usually 298 K and 100 kPa
stoichiometric coefficient	the number of moles of a species as shown in a balanced equation
square planar	the spatial arrangement of a central atom surrounded by four atoms situated at the corners of a square
tetrahedral	the spatial arrangement with one atom at the centre of a tetrahedron of four other atoms
trigonal bipyramidal	the spatial arrangement with one atom at the centre of five other atoms, with three atoms in its plane, one atom above this plane and one atom below this plane
trigonal planar	the spatial arrangement with one atom at the centre of a triangle of three other atoms, with all four atoms in the same plane
van der Waals' forces	the weakest forces of attraction that exist between atoms or molecules. These forces result when electrons on adjacent atoms are displaced and induce temporary dipoles (also known as dispersion forces or London forces)

Index

Notes

Notes

Notes

Notes

Notes